Legal Meanings

Foundations in Language and Law

Editors
Janet Giltrow
Dieter Stein

Volume 1

Legal Meanings

The Making and Use of Meaning in Legal Reasoning

Edited by
Janet Giltrow
Frances Olsen
Donato Mancini

DE GRUYTER
MOUTON

ISBN 978-3-11-126610-7
e-ISBN (PDF) 978-3-11-072096-9
e-ISBN (EPUB) 978-3-11-072100-3
ISSN 2627-3950

Library of Congress Control Number: 2021938816

Bibliographic information published by the Deutsche Nationalbibliothek
The Deutsche Nationalbibliothek lists this publication in the Deutsche Nationalbibliografie;
detailed bibliographic data are available on the Internet at http://dnb.dnb.de.

© 2023 Walter de Gruyter GmbH, Berlin/Boston
This volume is text- and page-identical with the hardback published in 2021.
Cover: kokouu/E+/Getty Images
Typesetting: Integra Software Services Pvt. Ltd.
Printing and binding: CPI books, GmbH, Leck

www.degruyter.com

Contents

Janet Giltrow
Legal meanings: Introduction —— 1

Anna Arzoumanov
Freedom of art in French legal proceedings: A discourse analysis perspective —— 17

Victoria Guillén Nieto
"What else can you do to pass . . . ?": A pragmatics-based approach to *quid pro quo* **sexual harassment —— 31**

Stanisław Goźdź-Roszkowski
Hostility to religion or protection against discrimination? Evaluation and argument in a case of conflicting principles —— 57

James Vanden Bosch
Heller **(2008) and the language of the Second Amendment: Grammar, meaning, and canonical conventions —— 77**

Jacob Livingston Slosser
Experimental legal linguistics: A research agenda —— 107

Jennifer Smolka and Benedikt Pirker
Pragmatics and the interpretation of international law: Two Relevance Theory-based approaches —— 131

Svetlana V. Vlasenko
Temporal meanings in legal translation: English-Russian lacunas and associated semantic uncertainties —— 165

Subject index —— 191

Janet Giltrow
Legal meanings: Introduction

"Pendez les blancs"

In "Freedom of art in French legal proceedings: A discourse analysis perspective," Arzoumanov analyses a corpus of cases in which two legal protections were in conflict: protection of Freedom of Art, and protection of privacy – that is, protection of persons or groups from injurious, stigmatizing, or hate-inducing expression. Arzoumanov's analysis offers an alternative to the characteristically "polemical" register of debate. Advocates for the freedom of artistic expression, and those for protection of privacy, often repeat emphatic claims without advancing beyond those claims.

Among the tasks for the court in Freedom of Art cases is to decide the aesthetic status of the expression in question – literary, musical, visual art, performance, or other – and to assign a relative weight to its protection, against the weight assigned to the protection of aggrieved parties. What counts as *art*? Does the instance in question fit the definition of art? For legal purposes, what does the instance mean? What is being said in this instance? Does the very thing it says support or undermine its definition as art? Such questions are difficult, even for experts in the aesthetic disciplines – literature, art history, musicology, and so on. In court, the questions are assigned to the bench. Widely reported, they also go out to the consideration of a general public, whose knowledge of art and its meanings stretch from folk to cognisant.[1]

In the French legal system, these questions have come to be addressed through the concept of *distanciation*. In deliberations and in pleadings, "we notice the predominant use of the French language term *distanciation*" (Arzoumanov 22). If English speakers seek a translation of this term, they will find no equivalent in English,

---

[1] In "The Maplethorpe obscenity trial," Mezibov (1992) describes a painstakingly formulated and executed strategy for bridging – rather than extending – this spread between folk and cognisant perspectives on visual art. With colleagues, Mezibov defended the curators of Cincinnati's Contemporary Art Center against charges of obscenity in exhibiting erotic photographs by Robert Maplethorpe.

Note: The chapters of *Legal Meanings* develop presentations from the September 2019 UCLA Conference of the International Law and Language Association, organised and chaired by Professor Frances Olsen, University of California Los Angeles, School of Law.

Janet Giltrow, University of British Columbia

https://doi.org/10.1515/9783110720969-001

and for the word's meaning they have to be satisfied with an impression rather than a definition. But in this, English speakers will not be at a disadvantage, for, despite its frequent use in creating legal meaning of art and art works in France, *distanciation* "is never defined, [the term] has a very imprecise scope and is used to refer to very different phenomena. Although it is often used in considerations, nobody has determined criteria of *distanciation*" (Arzoumanov 22). Meanwhile, the relevant genres are "mostly humour, satire, caricature, rap and fiction. Lawyers assume that these imply an inherent *distanciation* and a specific interpreting pact." (Arzoumanov 23). When a controversial work is found to be in one of these genres, it can (but not necessarily does) get defined as *art*, and therefore deserving of protected status (Arzoumanov 17). The question for pragmatics is: Is it *genre* itself, (i.e., the role that *distanciation* plays in an "interpretive pact") that deserves protection? Or is it the particular genre (e.g., rap songs but not the job interview)? Is it the particular genre or the instance of that genre (e.g., humour but not *this* joke)?

In having identifiable features, rap music (today known more widely in English as *hip hop*) bids for *distanciation* via genre. Arzoumanov summarizes genre features as they show up in arguments to establish a legal meaning for *rap*:

> the rapper is the spokesman for the unhappiness felt by the young generation [. . .] rap involves verbal abuse and raw lyrics, a style that has its own code of reception [Rap] is linked to the right of a minority to express itself. Rap music voices their aspiration to be included in society. (Arzoumanov 23)

The court's reasoning in four cases which involved controversial rap performances testing the genre's legitimate *distanciation* is then examined.

In the first case, rapper Orelsan was prosecuted by "several feminist associations for his violent lyrics," these lyrics said to "[incite] hatred against women" (Arzoumanov 25). In the first instance, "the judges decided that the meaning of the words should be weighted more heavily than the context" (Arzoumanov 26). On appeal, the decisions were "the exact opposite . . . even though [the judges] examined the same words" (Arzoumanov 26). Possibly, *distanciation* hovers around traditional (but equally intuitive) distinctions between literal and non-literal meaning. Both the literal/non-literal distinction and the *distanciation* criterion produce meanings useful to legal reasoning.

Arzoumanov's fourth exemplary case may test this possibility. Rapper "Nick Conrad was prosecuted in 2018 for criminal incitement against French white people and convicted" (Arzoumanov 27). Lyrics in Conrad's song "Pendez les blancs" [Hang white people] were judged to be incitement to violence. (For example, Arzoumanov provides excerpts from Conrad's song: "Je rentre dans les crèches, je tue des bébés blancs, [. . .] pendez leurs parents" [I go into the nurseries, I kill white babies, [. . .] hang their parents] (Arzoumanov 28)). Neither the claim of adhering to genre conventions, nor the claim that the lyrics' meaning was a "[denunciation of] racism against [the rapper's] community" (Arzoumanov 28), tipped the balance for freedom of art. The court took the literal meaning, and put aside meanings mediated by *distanciation*.

"Literal meaning" is a prime in public consciousness. Where is the common-sense in *what he said is not what he said*, or *what he said is not what he meant*? It might help (or might not) to turn the claim slightly, to *he is saying what is said / he is saying what others have said*, or *he is saying what others typically say*. Rather than saying these words, he is *showing* these words to his audience, and leaving the audience to make inferences. You can't infer definitively that the meaning *denunciation of racism* is the intention behind showing these words. And how can one know that this, and not the literal meaning, is what is intended? The analysis might not convince people, including judges. They might listen to demonstrations that this indirect way of getting a meaning across is common in everyday interaction. But then, once we know that this is commonplace in conversation and many other kinds of talking, how do we know that this indirection has an aesthetic motive, that *distanciation* is what is intended?

Offer . . . a hint . . . or mere hint

Arzoumanov's study of courts' finding – or missing – aesthetic meanings in rappers' lyrics is a study in indirect meaning, or indirect communication. In many cases, the rappers do not say, explicitly, and in so many words, what is meant. Indeed, it's hard to think of any instance of *art* which does not intend its meaning indirectly rather than directly. Long before *Relevance: Communication and cognition* (Sperber & Wilson 1995 [1986]), Dan Sperber, in *Rethinking symbolism* (1974), asks why go to the "[disproportionate]" trouble and "exorbitant" expense of myth or ritual when the meaning could be stated directly (7–8)? Where is the profit in this? Seemingly, even as it is more demanding and uncertain, indirect speech repays the cost by opening meaning to more possibilities.

Grice and those who followed went to many instances of indirection beyond those found in art and literature. Most came from examples invented from everyday life, and from exchanges on commonplace topics. None of these cases is about meaning being hidden, coded, then discovered only in close reading or with a key. In every case, the meaning is available to the hearer, although probably not so accessible to an overhearer.

Victoria Guillén Nieto's study of a case of verbal sexual harassment is one of these instances. A Chinese post-graduate student at a Spanish University alleged that a professor had engaged in *quid pro quo* sexual harassment while she was writing an exam in his office. That is to say, the professor had intimated a sexual bargain in return for a passing grade in the course. As Guillén Nieto writes, the "student appended to her complaint a 20-minute audio recording she had managed to make of the incident" (Guillén Nieto 37). This documentary evidence took the case out of the category of allegedly unsubstantiated recall of the literal *what was said*, in so many words. Even so, the first panel of adjudicators did not agree in their interpretation of the professor's remarks: "the three male professors concluded that the complainant had not been a victim of sexual harassment, the three female professors concluded that she had been" (Guillén Nieto 37). (Notably, two years later, with an improved transcript of the recording, the committee reached a unanimous decision supporting the student's claim.)

Just as in the cases adjudicating rappers' lyrics, a search for "ordinary meaning" cannot take the case beyond such an impasse. Like French adjudicators in the Freedom of Art cases not hearing aesthetic meaning, the male panellists who could not hear harassment, in effect, took the meaning of the remarks from what was directly said. The other three panellists heard what was indirectly said. Both the rapper and the professor have reason to resort to indirection: for the rapper, there was aesthetic benefit – a surplus of meaning. For the professor, there was benefit too, but not aesthetic. Nieto cites Scarduzio and Geist-Martin:

> *Quid pro quo* sexual harassment falls into the category of sexual coercion, [. . .] it refers to situations in the workplace in which individuals who hold power offer, or merely hint, that they will provide advantages or withhold disadvantages in return for the target's satisfying of a sexual demand. (Scarduzio & Geist-Martin 2010, cited by Guillén Nieto 31)

An *offer* . . . a *hint* . . . or *mere hint* – *quid pro quo* sexual harassment may actually be identified, rather than cloaked, by its indirect expression.

The case in the first instance being stuck on direct/indirect, Guillén Nieto offers models from linguistics. These provide different terms for analysing and understanding the interaction, in a "top-down pattern of analysis." At the top is Hymes' (1974) Speech Event, and related models (e.g., *schema*); midway, and leading to micro-analysis (e.g., Conversation Analysis, Speech Act analysis), is Brown and Levinson's (1987) Politeness Theory. A Speech Event can be defined by an observer's detecting and defining components and their order. Guillén Nieto composes a list of features for the Speech Event of "an exam": "We can then predict that "an exam" as a speech event is bound to a speech situation, including [. . .] typified contextual elements." For the Speech Event "an exam" these elements include, for example

> (e) Mode of communication. The mode of communication may be oral or written. In either case, one expects the participants to use a formal register and a respectful tone.
>
> f) Standards of conduct. One expects professors to watch the exam, in order to guarantee that all students have the same opportunities to answer the questions. The students are aware of being under surveillance by the professor. Both professors and students must be civil and respectful to each other. (Guillén Nieto 40)

But Guillén Nieto knows that the analysis of components misses participants' experience of them, and of the Speech Event. Accordingly, students were invited to participate in a structured prompt-and-response survey. Sixty-five students were presented with a facsimile of the situation under investigation – the student/professor encounter. They were then asked to pick out which elements of the Speech Event departed from expectation of what goes on at an exam. There was a high degree of agreement (90% and higher) on which elements were sites of unexpected behaviour (Guillén Nieto 50). The students agreed on what was going wrong.

Guillén Nieto also exemplifies application of Politeness models to passages of interaction. The model exposes face-threatening acts, the effects of which could include feelings of embarrassment and intimidation. Like Speech Event analysis, Politeness analysis can make intentions in indirect communication audible to overhearers. For legal linguistics, the question is: can these analyses produce meanings which can be useful in legal reasoning? At the very least, they contribute resources for finding indirectly communicated meaning. At most, they could take the professor/student interaction to the same level of interpretation of indirect meaning as some courts reached in Arzoumanov's study of legal reasoning about Freedom of Art.

"Sincerely held religious beliefs"

As in the cases studied by Arzoumanov, that studied by Goźdź-Roszkowski finds two areas of protected rights in contention. Under the First Amendment to the US constitution and the Fourteenth Amendment, opposing stances may contend.

> First Amendment: Congress shall make no law respecting an establishment of religion, or prohibiting the free exercise thereof; or abridging the freedom of speech, or of the press; or the right of the people peaceably to assemble, and to petition the Government for a redress of grievances.[2]
>
> Fourteenth Amendment: All persons born or naturalized in the United States, and subject to the jurisdiction thereof, are citizens of the United States and of the state wherein they reside. No state shall make or enforce any law which shall abridge the privileges or immunities of citizens of the United States; nor shall any state deprive any person of life, liberty, or property, without due process of law; nor deny to any person within its jurisdiction the equal protection of the laws.[3]

Same-sex marriage was legalized in the United States under the "equal protection" provision in the Fourteenth Amendment. Principles derived from these amendments contended in *Masterpiece Cakeshop, Ltd., et al. v. Colorado Civil Rights Commission et al.*,[4] a case so striking that it came to attention outside the United States. International publicity was driven not simply by the abstractions of the constitutional provisions that were at stake, but by the instance which brought these abstractions to life and into conflict between social factions: "different social groups interested in the outcome of specific civil rights cases" (Goźdź-Roszkowski 57). When a decision is needed to resolve conflict between two constitutional protections, how does a court show – and justify – its reasoning? In this case, the "Supreme Court of the United States in 2018 ruled in favour of a Colorado baker who had refused to create a wedding cake for a gay couple" (Goźdź-Roszkowski 58).

Outside the United States, the notorious aspect of the decision was the role of cake in legal reasoning. The first constitutional claim raised by the baker, and cited in the decision, was that requiring him "to create a cake for a same-sex wedding would violate his First Amendment right to free speech by compelling him to exercise his artistic talents to express a message with which he disagreed" (Goźdź-Roszkowski 69). In this role, the meaning of *cake* includes *means of expression*. This is a sense of *cake* which departs from ordinary meaning. In its reasons, the court "attributes" rather than "avers [. . .] these conflicting assessments": cake as art, cake as product of skilled labour (Goźdź-Roszkowski 69). Court's Reasons keep the balance, neither protection outweighing the other, neither tipping the balance in a "difficult" case (Goźdź-Roszkowski 71).

---

[2] https://www.law.cornell.edu/constitution/first_amendment (accessed 23 April 2021).
[3] https://www.law.cornell.edu/constitution/amendmentxiv (accessed 23 April 2021).
[4] https://www.oyez.org/cases/2017/16-111 (accessed 23 April 2021).

The court's attention is drawn to the reasons in the decision which is appealed. The decision from the Colorado Commission of Human Rights gives greater weight to the protection of rights to accommodation – greater weight, that is, to people's rights to, for example, retail purchase, rental, education, health care, protection of their "privacy" by prohibiting, even at the shop door, discriminatory practices which could stigmatise entire sections of the population. However, in the USSC judgement, as Goźdź-Roszkowski shows, the contending principles are kept formally in balance in the USSC judgement itself.[5]

> The Civil Rights Commission's treatment of his case has some elements of a clear and impermissible hostility toward the sincere religious beliefs that motivated his objection.
> (para 14, cited in Goźdź-Roszkowski 66)

In the judgement, <cake> does not have to defend itself as speech. The temporary departure of <cake> from ordinary meaning (presumably, neither dictionary nor corpus nor common-sense would confirm this meaning of cake) provides material for the court's rationalisation of its judgement. It is a legal meaning insofar as it offers this material to legal reasoning. Yet, as Goźdź-Roszkowski reports, the rationale does not depend on it. Meanwhile, no card has been played in the game balancing protected rights.

Grammar v. grammar

James Vanden Bosch consults corpora broadly and specifically, and in their versions over time, to establish the authority of grammatical analyses in reading the Second Amendment of the US Constitution: "A well regulated Militia being necessary to the security of a free State, the right of the people to keep and bear Arms, shall not be infringed." Famous in America, this amendment to the Constitution either guarantees citizens' rights to "keep and bear" firearms – or, on the other hand, it limits these rights to the narrow membership of a "well regulated Militia."

The Second Amendment has been read many times. When in 2008 it was read by Justice Scalia in *District of Columbia v. Heller*, "firearms" had already been subjected to scrutiny for its plain and original meaning, corpus study helping out with that reading. The structure of the sentence and its ordinary meaning had also been subjected to legal grammar. Legal grammar analyses the sentence

---

[5] But, nonetheless, Phillips was entitled to the neutral and respectful consideration of his claims in all the circumstances of the case.

as having a *prefatory* element followed by its *operative* element. While the prefatory element is sometimes assigned "purpose," it is overshadowed by the operative clause. It cannot restrict or extend or confirm the meaning of the operative clause. As Vanden Bosch shows, according to legal grammar in both terms and function, the operative clause dominates the prefatory clause, taking no direction from it.

We know that "grammar" is not the speaking and hearing, writing and reading that we do. Grammar is an instrument for studying that speaking, reading, and hearing. It describes structures, and there are more ways than one for structures to be observed or brought to light.[6] Accordingly, there is not one grammar, but possibly, many grammars. Usually these grammars keep their distance from each other. In bringing to light legal grammar's competition with traditional grammar in analysis of the Second Amendment, Vanden Bosch gives us an opportunity to look into a case where different grammars come up against each other.

Vanden Bosch's "*Heller* (2008) and the language of the second amendment: grammar, meaning, and canonical conventions" demonstrates first that the prefatory clause in the Second Amendment is an "absolute" in traditional grammar (or "non-finite," its verb having no tense), the operative clause is the main (or finite) clause. The legal grammar at work here can supply rules of construction: in this case, the role and interpretation of "A well regulated Militia being necessary to the security of a free State." It can identify parts of the sentence in such a way as to make them visible, not only to the interpreter but also to those who read the interpreter's rationalisation of the judgement. Vanden Bosch's study also demonstrates that Justice Scalia was not wrong in his analysis of this sentence in its parts. A grammar is not true or otherwise; it is more or less coherent and serviceable.

What Vanden Bosch's study does demonstrate about Justice Scalia's application is, first, that the legal grammar may not, in this case, answer well to people's experience. It is hard to hear the absolute or prefatory clause as having no relation, or as having a negligible relation, to the finite or operative clause, or as having only such relation as can be disregarded or downgraded, in a provision of a constitution.

As shown by Vanden Bosch's corpus review of the legal absolute, it can advance not only *purpose,* but also, as here, prevailing circumstances, the statement of which could be understood as having only ceremonial relevance – or a heartier relevance as an expression of authority. The question, then: is the Second

---

6 Arzoumanov's (22) use of studies in "folk linguistics" is one direction that could be taken. And, halfway between traditional and folk, there are grammars for learners of additional languages.

Amendment's use of this structure the ordinary use, the plain use of this grammatical structure? Or is it a special one?

In answer, Vanden Bosch turns to corpora. As corpora can be consulted to find spans of word use and meanings contemporary to the document in question, Vanden Bosch uses data from the Corpus of Early Modern English and from the Corpus of Founding Era American English to demonstrate the work of the *being*-absolute in more or less formal 18th-century American documents. In most instances cited, the absolutes eligible to be analysed in legal grammar as "prefatory" are so weighty and substantial as to relegate the legal-grammatical "operative" clause to not much more than a wrap-up, the absolute having made it nearly inevitable. For example, as cited by Vanden Bosch:

> 1781: Order in council concerning the copying of public records, [10 May 1781]; Virginia executive council:
>
> The letters and other Papers of the Council having been destroyed in the expedition of the enemy to the Town of Richmond in the month of January last, and **it being of general importance that memorials of public events be preserved**, and particularly interesting to those having a share in the administration that the records of their proceedings should under every possible circumstance guard them against misrepresentation and mistake and the board being of opinion that copies may be obtained of many letters and other papers of considerable importance by application to those to and from whom they have been written, Advise that a proper person be appointed to execute this business, that he be instructed particularly to go to Congress and to General Washington in order to obtain permission to copy the letters which have passed between them and this board previous to the commencement of the present year (Vanden Bosch 79)

Meanwhile, as Vanden Bosch points out, legal and public reasoning focussed, perhaps more readily, on 18th-century meanings of *bear arms*. Justice Scalia cited "our review of founding-era sources" in establishing from these sources the 'natural' meaning of *bear arms*, as unambiguously used to refer to the carrying of weapons outside of an organized militia." This discovery which Scalia declares in effect proved the legal-grammatical status of the absolute: it is merely prefatory and contributes no substance to the operative clause (called even in traditional grammar "main" or "independent" clause, and, with these labels, putting the "prefatory" structure in its humble place). Vanden Bosch quotes expert, corpus-derived refutations of Justice Scalia's analysis (Vanden Bosch 81), and follows up by his own consulting of the Corpus of Early Modern English, and the Corpus of Founding Era American English, searching for *being necessary* in sentence-initial, medial, final, and cleft positions. Vanden Bosch's findings, along with those of other authorities, suggest that the sentence-initial absolute was versatile (and finding the *being necessary* absolute more common in America than in Britain).

Vanden Bosch's search for evidence is compelling – particularly in his foregoing the finding that Scalia's opinion is wrong in its relying on legal grammar rather than the resources of traditional grammar, and in its being out of earshot of "natural" readings. He gathers more evidence, and then more, while he reviews evidence already in-hand, puts every <being necessary> in a setting beyond minimal co-text. The study is a high-value use of corpora for law.

Floating hotel or seagoing train?

Slosser proposes that Conceptual Metaphor Theory be the basis for this agenda for experimental inquiry. Practically, the idea would depend on first getting rid of people's schoolroom ideas: namely, the idea that metaphor in its flair and flippancy would not be good for sober reasoning. Yet metaphor has long been accepted and studied as occurring widely, across solemn interactions and informal ones – and beyond poems. Metaphor has shown up in classic examples in pragmatics: "You are the cream in my coffee" (Grice 1989: 34).

Possibly, its built-in capacity for a functional indeterminacy has made people say that, at least, metaphor should be used only sparingly. In any case, the Conceptual Metaphor is to be found on a plane different from, although not unconnected to, the plane of, for example, coffee-cream relationships.

In proposing a Conceptual Metaphor-based methodology, Slosser notices the tendency to take the language of judgements as "proxies" for judicial reasoning, and the tendency to look for the reasoning by collecting features of the language of judgements. Slosser questions whether these collections are usable. From the view of cognitive linguistics, metaphor "is basic to thought itself" (Slosser 109). Rather than the surface "flourish" of intentionally figurative language, "it is the more hidden and implicit presence of metaphor in language that offers much for the legal linguist to consider. From the perspective of Conceptual Metaphor Theory, metaphor is a cognitive mechanism that structures thought" (Slosser 109). Conceptual Metaphor, then, finds language involved in reasoning more fundamental than argument, and more motivating than persuasion. For (my adapted) example, a shared but unexpressed understanding of *high* and *low* supports terms/interpretations like *flighty, superficial* vs. *profound, deeply reasoned*. Or *soar, illuminate* vs. *lifting from, rising from destitution, bury*. Metaphor on this plane is not just one speaker's metaphoric representation, and not just a presentation affecting or aiming to affect reasoning. Metaphor is an unbidden current from one thinker to the next, not just one speaker's presumed "word choice," or "choice" of a metaphor just like someone else's. And at this level, metaphor is not inducement to reasoning, but is reasoning itself.

In one study of this phenomenon,

> participants – investors, in this case – were asked whether they thought a trend in the stock market would continue, after having been given a description of the market along with some graphs. Participants were asked to read texts framed with metaphoric phrases describing an agentive market trend, like "jumped," "climbed," "wandered," while others read materials using non-agentive terms like "swept" or "plummeted." (Slosser 112)

Those who read the first selection, with an agentive framing, predicted that the market trend would continue.

Slosser's earlier work on conceptual metaphor in legal reasoning found metaphoric congruency between the winning argument, and the judge's reasons, congruency with precedent, and even congruency over a long chain of precedent. To get closer to this congruency, Slosser designs an experimental model based on the well-known example of the *Adams v. New Jersey Steamboat Co* case. A passenger loses his suitcase on the ferry, and wants to be compensated. One precedent would find analogy between ferry and hotel, another between ferry and train, with different implications for liability. In Slosser's experiment, one precedent will be "framed" to "fit" with *train*, the other to fit with *hotel*. Conceptual Metaphor Theory suggests that it is worth finding out if "metaphoric fit" matters as much or more than facts and law. The experiment's design is spelled out in the chapter, one of its virtues being its replicability.

Slosser's work extends and complicates the role of language in legal meaning. What is in mind here is not language itself delivering legal meaning. Instead, wordings activate a kind of recognition, which in turn puts some meanings in reach, and not others. In Slosser's work, language is not a script for the strategic uses people often attribute to writing – in so many words, it is not strategic.

That is to say...

Like rules for interpretation in other legal genres, canons of interpretation of international treaties call for "ordinary meaning." In the search for ordinary meaning, Smolka and Pirker point to "intended meaning." Since the treaties themselves signal consensus about "what is intended," pragmatic analysis will significantly illuminate what that intended meaning is, in the treaty (Smolka & Pirker 132). ("Ordinary meaning" in this case might be something like "shared meaning.")

Guided by Ariel's (2016) version of Relevance Theory, Smolka and Pirker go with her baseline "explicated inference." Explicated inference is a practical rather than abstract definition of explicature insofar as it can be coaxed out by *that is to say* (as below) (Smolka & Pirker 140). Smolka and Pirker also adopt Ariel's "discoursal status": the relative prominence of the inference in the range of what is being talked about. Discoursal status lines up (I think) with Grice's (1989) Intended Meaning. If you don't get the implicature, you have missed the point, the point being an element that must have high discoursal status.

In the classic "I am tired" example, in Sperber and Wilson (1995), the high-status topic is not the *osso buco*, but Peter's implicated refusal to prepare this complicated dish after he's had a long workday.

> Mary: What I would like to eat tonight is an osso buco.
>
> Peter: I had a long day. I am tired. (146)

Once the fact that Peter is not going to cook is inferred by *that is to say*, it takes a higher discourse status than osso buco. No one is going to keep talking about osso buco.

Most compelling in Ariel's typology are the practical instruments it offers, and their basis in theory. Smolka and Pirker pick up the re-wording instrument. They use it on working examples from the interpretation of international treaties. To explicate inference, re-wordings summon candidates for ordinary meaning, and then test those candidates. For example, Ariel offers a way to test pragmatic enrichment, by rewording "what was said" using the "that is" test. As Smolka and Pirker explain, "One adds a 'that is (to say)' clause to spell out the explicated inference. If the pragmatic interpretation is an explicature, a correct utterance is the result" (Smolka & Pirker 142). Smolka and Pirker give the following examples, drawn from Ariel:[7]

> **Original:** My son said that *she* wasn't the last *one*. We're waiting for the next *one*. The speaker's son said that *Busaina Abu Ghanem* wasn't the last *female murder victim in the family*. We're waiting for the next *female murder victim in the family*.
>
> **Re-wording:** The speaker's son said that she, *that is (to say) Busaina Abu Ghanem* wasn't the last one, *that is (to say) the last female murder victim in the family*. We're waiting for the next one, *that is (to say) the next female murder victim in the family*. (142)

---

[7] Nearly all Ariel's examples are excerpted from a news article on "family honour" killings. "A Ramle resident: 'How can we talk against the murder of women? The next day they will riddle our house with bullets'" (Originally Hebrew, *Haaretz*, 2 November 2014, cited in Ariel 2016: 6).

Smolka and Pirker[8] apply the "that-is" test to the court's interpretation, and we find that the court's interpretation is an explicature – and importantly for international lawyers – that interpretation has, therefore, a prominent discoursal status:

Original: Once that period has lapsed, *the two parties* may submit the matter to the International Court of Justice.

The court's interpretation: *Either one of the parties* may submit the matter to the International Court of Justice.

Test: Once that period has lapsed, the two parties, *that is (to say) either one of the parties*, may submit the matter to the International Court of Justice.

from Ariel, cited by Smolka and Pirker, re-reading for strong implicature and requiring more re-wording than *that-is, or what-is-said*:

R_1: And John Doe, who is a company director, pretends to know that the balance sheet is going to be good so he starts buying.
S: OK that's a criminal offence.
R_2: Eh . . .
S: It's a bit of a criminal offence.
R_3: So he has a mother-in-law.
S: For this you go to jail.

"The speaker literally said that John Doe has a mother-in-law, but he actually indirectly conveyed that John Doe would illegally buy shares under his mother-in-law's name."

From Smolka and Pirker:

Original: The Court of Justice shall review the lawfulness of acts other than recommendations or opinions of the Council and the Commission.

The Court's interpretation: The Court of Justice shall review the lawfulness of acts other than recommendations or opinions of the Council, the Commission *and the Parliament*.

Test: The treaty *literally* stated that the Court of Justice shall review the lawfulness of acts other than recommendations or opinions of the Council and the Commission, *but it actually indirectly conveyed that* the Court of Justice shall review all acts of all EU institutions with binding legal effects towards third parties.

The advantage of describing the court's approach in these terms is that it becomes clearer that the court is not simply inventing elements that were not laid down in

---

[8] Smolka and Pirker's examples come from documents related to international treaties.

the treaty, but is replacing existing elements with – in the court's view – other, more plausible elements, or giving these elements a more pragmatically plausible interpretation. This does not mean that, legally speaking, the court is necessarily "right." But it focuses the discussion on whether the replacement is justified, rather than only the "invention" of new elements (Smolka & Pirker 145).

Pragmatics has gone ahead on intuitive listening, hearing ordinary interpretation of examples, to test, refine; discover principles and make new examples – a philosophical method, that here becomes an applied method, by means of Ariel's re-wording instruments. Smolka and Pirker's intrepid application also gets back to first-hand, practical contact with originating principles. Re-wordings should speak for themselves, without linguistic explanation. If they don't, their standing as "ordinary meaning" would be in question.

A short *dekada* and a long *sutki*

In "Temporal meanings in legal translation: English-Russian lacunas and associated semantic uncertainties," Svetlana Vlasenko looks into those lacunas. Roughly (and from an amateur point of view), translation pairs range from (a) we have a word for that; (b) we don't have a word for that but this word comes close; (c) we don't have a word for that but these words, this phrase gives an idea of that; (d) we don't have a concept of that. In (c) the lacuna is semantic; in (d) it is conceptual. Where would you put the Russian temporal expression *dekada* in this scheme? Derived from the same Greek root, the English word *decade* seems a promising candidate to correspond with *dekada*. But *dekada* is (at core) *ten days*. As Vlasenko writes, it is a a challenge to find wording in Russian for the English *decade* that would fit the name of UNESCO's endeavour, the *World Decade of Cultural Development*, which was slated to last from 1988–1997 (Vlasenko 185). The mis-fit could be attributed to "culture," but by Vlasenko's methods "culture" is rendered as shared, culture- and language-specific "experience of time" (Vlasenko 166). The word, or words, found when *there is no word for it*, or when *there is no concept*, has to be "comfortable" to Russian lawyers to be a successful legal translation (Vlasenko 172), a translation that can be used in legal reasoning.

In estimating degrees of correspondence, Vlasenko draws the horizon along which temporality is known and marked. In the process, clusters of English, as well as Russian, words appear. They come up not simply as signs of timing, but as expressions of a shared experience of temporality. These groups of English words can refresh English speakers' shared markers for their own experience of time. Consider, for example: *whatever the case may be / ex-wife, ex-husband / the then-president / from time to time / at all times / lapse of time / no later than / on or before / early / until recently / undated* (Vlasenko 168–9). Rather than consult authority for definitive translations – which after all might not function well in the new context – Vlasenko's method lines up the English examples across from Russian ones, finding fit partners or not, and negotiating the mis-fits. This methodology gives the translator more to work with than the sought-after one-to-one correspondence.

In legal documents, temporal expressions bring the document into effect, at once or over a period; they bring the effect to a conclusion; they limit or extend the effect, or leave it to expire. They sequence roles, and dictate the consequences of actions. As speakers' shared experiences, they are an important focus for translation studies. Temporal wordings express (as well as consolidate) shared experiences of, as Vlasenko notes, a universal phenomenon – namely, *time*. Vlasenko's aligning of English terms with interests sympathetic to expressions in Russian show that those experiences are not always ready to be shared across translation, no matter the universality of *time*.

After *dekada*, the Russian term *sutki* is the object of Vlasenko's most extensive analysis. Entries in the Russian column are numerous. Correspondences could suggest that the meaning of *sutki* is close by and easily accessible: *24 hours, 24-hour period*. The amateur translator, or one not concerned with specifically legal meanings, might be comfortable a bit longer, even when they hear about *sutki* extending beyond this timing to temporal entities associated with different fields, such as "labour, transport, medical" and different "conceptual perception."

> conceptual perception of *sutki* as continuous temporal spaces, burdened with time-tabled arrangements and/or practices, which fall under various regulations. A 24-hour period is treated differently in various fields of law. Essentially, this period may begin at the birth of a stillborn baby or at the sudden death of an employee at the workplace and, as such, it may be indicated in relevant legal documents [. . .] This concept of *sutki* is specifically important, as it is rarely tied to the 00:00 hour. [. . .] Some Russian regulations name a daytime portion within the 24-hour period as *dnevnoje vremya sutok*, while naming a night-time portion as *nočnoje vremya sutok*. (Vlasenko 181)

Curfew, emergency, and enforcements: *sutki* can be regarded as a concept perceived by law-enforcement agencies as a timing unit with its length/duration equalling 24 hours, but with variable start-end properties if it accrues in situations of detention and confinement. *Sutki*'s occurrences are far-flung, found in one place and then another, after being captured briefly for a legal meaning. *Sutki* may ask for translation at each type of occurrence. But that may not suit its nature.

Equally, *dekada*, which is at first challenging but then manageable, has trouble in store. Just as *sutki* can seem to promise *24 hours*, *dekada* seems to promise 10 days. This promise comes with qualifications. It doesn't prevent the phenomenon of speakers of the paired languages hearing their language's own meaning – English speakers hear *decade* in *dekada*; Russian speakers *dekada* in *decade*. And, despite *dekada*'s being ready for, if not comfortable with, *10 years*, there is still "no word" in English for *dekada*'s meaning "a ten-day period devoted to an event or phenomenon of public significance, usually in literature or arts" (Krysin, qtd. in Vlasenko 183).

References

Ariel, Mira. 2016. Revisiting the typology of pragmatic interpretations. *Intercultural Pragmatics* 13 (1). 1–36.
Brown, Penelope & Stephen C. Levinson. 1987. *Politeness: Some universals in language usage*. Cambridge, UK: Cambridge University Press.
Grice, Paul. 1989. *Studies in the Way of Words*. Cambridge, MA: Harvard University Press.
Horn, Laurence. 1972. *On the semantic properties of logical operators in English*. Los Angeles: University of California PhD dissertation.
Hymes, Dell. 1974. *Foundations in sociolinguistics: An ethnographic approach*. Philadelphia: University of Pennsylvania Press.
Mezibov, Marc. 1992. The Maplethorpe obscenity trial. *Litigation* 18 (4). 12–15, 71.
Sperber, Dan. 1974. *Rethinking symbolism*. Cambridge, UK: Cambridge University Press.
Sperber, Dan & Deirdre Wilson. 1995 [1986]. *Relevance: Communication and cognition*. Cambridge, MA: Harvard University Press.

Anna Arzoumanov
Freedom of art in French legal proceedings: A discourse analysis perspective

Settled concepts in French law protect personal integrity and privacy and also freedom of art. Conflict between these competing promises of protection excites discussion and legal activity. One concept can protect minorities from outspoken expression of injurious stereotyping which puts person and property at risk. Meanwhile, minorities' vehement attacks on majority domination may be eligible for protection under freedom of art, because contemporary art defines itself, in part, in terms of an aesthetics of transgression and continues to play with the boundaries of what is ethically permissible. Minority or majority interest can resort to one or the other of these two legal protections. When an instance of public expression defies public sensibility, and freedom of art is called on to protect this instance, art itself can call for definition. How do we know that this public expression is art? What is art? Once 'art' has been found by the court, how is weight assigned across the competing protections? How do we know that a disparaging insult to a group is aesthetically conceived? Or that a call to violence is aesthetic rather than literal? How do we balance between freedom of art and respect of other individual freedoms, such as freedom of worship or protection of privacy? What do these conflicts reveal about the French public's cultural sensitivities and the evolution of artistic practices? The answers to such questions are often given by taking a polemical stand for one side or another. Some believe there must be an exception made for art, and complete freedom of art. Others believe that the protection of fundamental freedoms must be further taken into account by artists. These controversies show that freedom of art is a blurred territory that reveals social tensions but that holds a lot of stereotypes at the same time. It is simply not enough to say that art is above the law or that any attempt to limit artists should be interpreted as censorship at a time when artists are pushing the boundaries of art, especially in the sphere of ethics.

Each case poses a challenge which requires a useful response. Here, we assume that a better knowledge of art controversies will contribute to new arguments that ultimately will better protect freedom of art. These questions are a strand of public debate that has been very important in the last few years. They have also been debated by jurists, as some of these controversies were brought before courts. How do lawyers judge art? What are the limits that the law imposes on artistic

Anna Arzoumanov, L'Université Paris-Sorbonne

https://doi.org/10.1515/9783110720969-002

discourse? How often are artists prosecuted in France, and on which legal bases? We assume that analysing the current situation by a comprehensive review of legal proceedings, as has never been done before, is a necessary first step to contribute to the public debate. My recent research has been focusing on that matter with the intention of taking a systematic approach to these cases (Arzoumanov 2021, in press). This research has had three major goals: building analytical tools to measure this phenomenon of censorship as it becomes increasingly prominent; helping jurists, academics and stakeholders by providing them with a corpus that aggregates court judgements about freedom of art since 2000; understanding how the principle of freedom of art is used by courts while analysing its efficiency and scope. In order to explore the conflict between law and art, my original contribution crosses two main academic fields that are rarely in dialogue in France: discourse analysis and law. In this paper, I will narrow my scope and present my discourse analysis perspective on French legal proceedings about art in three parts. First, I will say a few words about the principle of freedom of art. Secondly, I will present the possible value of a discourse analysis perspective. Thirdly, I will focus on one category that is widely used by French courts, *distanciation*, in order to explore its efficacy in the debate.

1 The principle of freedom of art

Since the 19th century, art and literature have claimed their social autonomy (Bourdieu 1992; Bénichou 1973). This constitution of art as a specific field is well known from art historians, theorists and sociologists, as it has had a major influence on the way art and literature have been defined. Because of this presumed autonomy, artistic creation has been distinguished from other types of expression. It presupposes its own intentions and specific codes of reception based on aesthetic judgment. Any intrusion of other categories of judgement into the field of art, such as those relating to ethics, thus appears as a misunderstanding of this autonomy. This conception also entails placing artistic creation outside or above the law and making an artistic and literary exception. Any opposition between artistic norms and legal norms can therefore be perceived as a conflict that undermines the very values of the art world. This conception of art and literature as autonomous is largely weakened by contemporary artistic practices, as Bernard Endelman, Nathalie Heinich and Carole Talon-Hugon, for example, have shown (Endelman & Heinich 2002; Heinich 2014; Talon-Hugon 2019). But it is still a common idea and furthermore this principle is back in the spotlight as art controversies increase. In that context, artistic lawsuits are often considered censorship resulting from justice's short-sightedness and misunderstanding of artistic rules. Art is thus deemed to be entitled to greater protection than other discourses, which each judicial intervention of the law would ignore. Ernest Pinard became the symbol of this illegitimate censorship: he was the famous prosecutor in the trials of Charles Baudelaire and Gustave Flaubert, who were accused of offending public morals in their books *Les fleurs du mal* (1857) and *Madame Bovary* (1856). This principle of artistic exception (Artous-Bouvet 2012; Sapiro 2019) is strongly embedded in the popular imagination, and the consequence is that trials involving artists, writers or musicians make a negative impression and are the subject of an associative and militant watch that has been particularly active over the past twenty years. In France, the *Ligue des droits de l'homme* (LDH) [League of Human Rights] through its "Observatoire de la liberté de creation"[1] continues to call for extreme caution and to publish press releases denouncing what they consider a major threat to freedoms. At an international level, UNESCO has also warned about the worldwide situation and some international associations, such as Freemuse, have been created to promote and defend freedom of artistic expression, pointing out the fact that "unnecessary and illegitimate restrictions are often placed on fundamental rights and freedom of expression in times of uncertainty and securitisation."[2]

Instead of reiterating stakeholders' arguments, that art is free and must be protected, I aim to change the perspective and to map and measure the situation of freedom of art in France by focusing on its legal regulation. First, we can point out the fact that French law appears to be out of step with common sense. In its legislation it makes no distinctions between forms of expression and can therefore limit the rights of artists and writers, just as it can limit any other discourse. However, it does not appear to reject completely the argument of the autonomy of art, or at least the idea that it should benefit from a higher level of protection. In the legal community, we can report a rising reflection about freedom of art. Let me begin by saying a few words about the legal system of freedom of speech in France, which is very specific, to better understand the situation of freedom of art.

The French legal system is based on civil law, which means that it is based on legislation. There is a specific law concerning freedom of speech which has

---

[1] The "Observatoire de la liberté de creation" was created in 2003. It continues to publish releases on its website about what they consider as a threat for freedom of Art: https://www.ldh-france.org/sujet/observatoire-de-la-liberte-de-creation/ (accessed 25 April 2021).
[2] "Security, creativity, tolerance and their co-existence: the new agenda of European freedom of artistic expression," *Freemuse* report, 2020: http://www.live-dma.eu/wp-content/uploads/2020/01/SECURITY-CREATIVITY-TOLERANCE-AND-THEIR-CO-EXISTENCE.pdf (accessed 25 April 2021).

been applied in France since 1881, called "loi du 28 juillet 1881 sur la liberté de la presse"[3] [the law of July 28 1881 on the freedom of the press]. It rules freedom of speech in France in general and it is still applicable today, even if the types of speech have evolved considerably. The 1881 law does not distinguish between different types of speech. Every discourse can be prosecuted, whether it is a photograph, a song, a novel, a tweet, a political speech, an advertisement, or whatnot, and the law is the same for all citizens. Therefore, freedom of art is not specifically mentioned in the law. This law proclaims that speech is free, but at the same time, it provides limits, including two main types of press offences. First, it sets limits on violations of individual rights and provides criminal penalties for offences such as defamation and insult. Secondly, it regulates some aspects of hate speech, such as incitement of hate or violence or apologia for crime or racism. In addition, since the 1970s, many legal proceedings use another legislation, named privacy protection, which was included in the civil code in 1974 (article 9).

However, *freedom of art* (*Liberté de création* in French) has been an increasingly important principle since its appearance in jurisprudence. *Création* has a broad interpretation. It includes all types of artistic discourse (literary, visual, performing arts), but more generally the process of making all kinds of original works. Every legal proceeding in which artists are prosecuted is based on the principle that freedom of art must have a different protection than general freedom of speech. A new law based on this principle was introduced in France (*Loi sur la liberté de creation et du patrimoine*[4]) [Law for the freedom of creation and patrimony] in 2016. This law is pioneering in that it proclaims that "artistic creation is free" (article 1). For the first time in international law, it establishes artistic creation as a public good. In addition, it specifies that the "dissemination of artistic creation is free" to ensure greater public access to artistic works. Meanwhile, at a European level, freedom of art is not mentioned in the laws. But the same observation could be applied to the jurisprudence of the European Convention of Human Rights that has developed a specific principle of freedom of art and applies a specific framework in judging artworks. In a nutshell, freedom of art is a right created by case law. The rise of this principle involves three points: being able to identify art boundaries, judging artworks with an appropriate and specific framework, and developing specific categories of judgement.

---

3 Loi du 29 juillet 1881 sur la liberté de la presse.
4 Loi LLCAP: Loi Liberté de création, architecture, patrimoine.

2 Presentation of the corpus and method

In my opinion, studying how freedom of art is regulated by courts is a good way to combine the fields of law and linguistics, for at least two reasons. First, discourse analysis provides a framework for the methodical selection and comparison of judgements in cases involving freedom of art. Its bottom-up approach, based on detailed data, can be used to reliably map the legal aspects of freedom of art. My individual work does not seek to give a complete overview, since the matter is too broad, but my goal is to deliver concrete results by analysing a smaller corpus. I am focusing on the 17th chamber of the court of first instance in Paris in the past twenty years. This chamber is very specific, it is called in French "chambre de la presse" [Press Chamber], because it is there that most trials regarding freedom of speech and freedom of art take place. It is well-known internationally because it is the court that judged the much-publicized case of the *Charlie Hebdo* caricatures of Muhammad in 2007. Although it is a first instance court, many observers report that it has gained increasing authority and that its decisions are often upheld on appeal and cassation. It is important to underline that the present corpus was scattered and not easy to access. Some data were published in databases specialized in freedom of the press (*Legipresse*) or general law (doctrine.fr, Dalloz, Lexis Nexis, Extenso, Lamy). But many of them were not published at all. Their collection has been made possible only by direct contacts with court clerks, judges and attorneys. My second criterion is based on the fact that the defendant refers to himself as an artist or a writer. 46 cases meet these criteria with 26 literary works (25 novels and 1 theatre performance), 7 photographic artworks, 5 movies and 8 songs. This result is significant by itself. We can compare it to the large number of decisions of the 17th chamber in the last twenty years (around 9000). It means that artists' and writers' trials are proportionally very rare. But most of them have been highly publicized in the media. This can explain the feeling of increasing censorship of artists and writers. In addition, when possible, I also analyse attorneys' pleadings, when they have agreed to furnish me with this material. In these documents, we can observe the parties' argumentation.

Secondly, this *corpus* requires a linguistics perspective, because the matter that is debated at trial is a list of contentious extracts chosen by the prosecutor. Each party has to provide an analysis of utterances in order to convince the judge that he has the correct interpretation. These utterances can be a single word or a few lines. This analysis must be explained in the pleadings, in the decision and in the considerations held by courts. They have to refer to the law but also set out and clarify the interpretation of the contentious statements. For that reason, jurists engage in a form of discourse analysis themselves, with their own categories and criteria, which can be applied flexibly. Their two major criteria are the meaning of words and utterances themselves and the context (identification of the artist's intention, evaluation of prosecutor's harm, analysis of the publishing medium, the type of discourse, the artistic genre). The goal of the study is inspired by the field of folk linguistics (Niedzielski & Preston 2003; Dennis Paveau 2008; Achard-Bayle & Paveau 2008; Husson 2015), which is in turn inspired by folk theory and which studies linguistic analysis conducted by non-linguists. Those criteria are very close to the technical terms used in discourse analysis. In other words, there are strong links between law and discourse analysis, insofar as the folk extractions of words and passages resembles those of the prosecutor. But the terms used in the two disciplines are not the same and the criteria can vary. In addition, even if legal analysis of statements is based on law and jurisprudence, it also depends on judges' opinions and beliefs about language, which we can see through their grounds, their categories and through their approximation formulas (abstract nouns, approximative markers, and the like). For example, context is a very important criterion, but it is never rigorously defined and it varies depending on cases, judges and attorneys' grounds. Many debates also focus on the polyphony that is felt to be a key criterion (Arzoumanov & Latil 2017), but the category is never used by courts and their analyses are most often approximative and not defined by terms from linguistics.

3 The example of the category of *distanciation*

We call the methodological framework used in legal proceedings *the balance technique*. It is a casuistic method, which means that the judge in each case has to weigh up which rights are more affected: freedom of art or personal rights or community rights? One of the questions that my work asks is the following. Is there a specific framework of interpretation when the contentious statements belong to artistic works? As we know, judges are not allowed to base their decisions on their own opinions. How do they face the difficult question of deciding when freedom of art weighs more in the balance than other rights? Which criteria and categories do they use in their interpretation? In the considerations held by the courts and also in the pleadings, we notice the predominant use of the French language term *distanciation*. Even if the word doesn't exist in English, the French term is transparent enough to not require translation. It is a crucial category, on which a court decision can be based. Yet, it is never defined, has a very imprecise scope and is used to refer to very different phenomena. Although it is often used in considerations, nobody has determined criteria of *distanciation*. Actually, in many cases, the word is used without any clarification of the agent of the verb *distancer* [to distance] nor its object. Who is subjected to this *distanciation*? The speaker? The public? From what? The content? The reality? What creates this *distanciation*? At what level should the analysis be placed to measure it? My hypothesis is that the use of the category is intuitive but should be clarified through linguistic analysis. Otherwise there will continue to be a great deal of legal uncertainty.

3.1 The criteria of genre: the example of rap

In some decisions, *distanciation* is determined relative to the context, on the basis of the creative genre, since some genres are supposed by courts to imply a particular enunciating role for the speaker, a role that must be taken into account when interpreting utterances. These genres are mostly humour, satire, caricature, rap music and fiction. Lawyers assume that these imply an inherent *distanciation* and a specific interpreting pact. I will focus on the case of rap, in order to demonstrate that the idea that rap involves a *distanciation* depends on specific beliefs about the genre. Jurisprudence tends to look favourably upon the genre, at least since 2003 (Droin 2016; 2019). It is based on certain beliefs about the musical genre and its inherent *distanciation*, pertinent to the genre of rap as a whole, which are repeated in some judgments but which can vary. They include three different criteria:
- An enunciative criterion: the rapper is the spokesman for the unhappiness felt by the young generation. According to that opinion, enunciative polyphony is inherent to rap.
- A stylistic criterion: rap involves verbal abuse and raw lyrics, a style that has its own code of reception (which supposes that the public doesn't interpret it narrowly).
- A pragmatic criterion: rap music is linked to the concerns of suburban youth and is necessary as a mode of expression for them. It is linked to the right of a minority to express itself. Rap music voices their aspiration to be included in society, which means it has a positive effect on society. Defenders of rap derive inferences of a desire for "social inclusion" from their lyrics, while censorious listeners hear racial hatred and calls to violence.

The applicability of these criteria can be debated. Sociologists specializing in rap claim that rap is much more diverse. We can also point out the fact that pragmatic criteria are very approximative, since courts are never trying to measure the effects of rap on society in any other way than assuming those effects. That seems to be a paradox, since the charges themselves suppose a pragmatic interpretation of the lyrics (racial hatred; criminal incitement). In addition, the link made between the enunciator's characteristics (being a spokesperson) and the lesser effect of the speech has never been proven. But above all, the evaluation of this inherent *distanciation* varies depending on the courts, because there are no clear criteria to define it. For some courts, rap isn't exempt. The genre does not involve a systematic *distanciation*. There must be other criteria. To illustrate, we can take two examples of discharge and two examples of conviction.

3.1.1 The case of Sniper

In order to understand the principle of the rap genre and its inherent *distanciation*, let's look at the appeal decision about the rapper Sniper. Sniper was prosecuted in 2003 for inciting hatred toward the police and public officials. The complete lyrics of one of his tracks ("La France") were prosecuted. Here are some examples of lyrics from the track: "On n'est pas dupes, en plus on est tous chauds, pour mission exterminer les ministres et les fachos" ["We're not fooled, we're set for our mission to exterminate the ministers and fascists"] and "Frères je lance un appel, on est là pour tout niquer, leur laisser des traces et des séquelles avant de crever" ["Brothers, I make an appeal, we're here to fuck everything up, leave them traces and after-effects before dying"]. An appeal court interpreted these statements by themselves as incitation of hatred, but discharged Sniper by consideration that rap involves significant *distanciation*:

> If the song's content of La France" [. . .] is obviously made with violent words against the Republic, its representatives and the police, beyond which are the contentious lyrics, we have to remind that rap music is a musical genre rooted in the folk culture which has its origin in poverty and pain, rejection and bitterness, and that in the incriminating song this musical genre cultivates and takes up insult, crudeness and verbal violence by putting them in rhyme, it is a mode of expression shared by the song's author and its interpreters in order to express the suburban youth's affliction and unhappiness, their rejection of being forced to face situations that they experience as unfair and that they interpret as an exclusion and phrase their aspiration to have another way of life than the one that excludes them and marginalizes them.[5]

---

[5] CA Rouen, ch. Corr., 14 déc. 2005, procureur de la République près TGI Rouen c/ A. Karl Junior, Bachir B., Machouche B.; Selmi R.; M. Catenoix, prés.; M. Picquendar et Mme Bellamy-Chaline, cons.; Me Dominique Tricaud, av.: Juris-Data n° 2005–297833. In French: « Si la chanson intitulée « La France » [. . .] contient à l'évidence à l'égard de la République, de ses représentants et des fonctionnaires de police des termes violents parmi lesquels figurent les paroles visées à la prévention, il n'en demeure pas moins nécessaire de rappeler que le rap est un genre musical enraciné dans une culture populaire trouvant ses origines dans la misère et la souffrance, le rejet et le ressentiment et que dans la chanson incriminée ce genre musical,

All the beliefs about rap underlined above are gathered here and justify the discharge of the rapper. Genre is here the predominant criterion to decide whether there is a *distanciation*. However, the decision never examines concretely if this *distanciation* can be identified by the public in general or by the expected or likely audience.

3.1.2 The case of Orelsan

Let's take another example, that of how the *distanciation* associated with the rap genre contributed to the discharge of the rapper Orelsan, who was prosecuted for incitement of hatred against women. Orelsan was prosecuted by several feminist associations for his violent lyrics. Here are some extracts from the prosecuted lyrics:

> Les féministes me persécutent, me prennent pour Belzébuth
>
> Comme si c'était d'ma faute si les meufs c'est des putes
>
> Elles ont qu'à arrêter de d'se faire péter le u-c
>
> Et m'dire merci parce que j'les éduque, j'leur apprends des vrais trucs
>
> Des fois j'sais plus si j'suis misogyne ou si c'est ironique
>
> J'serai peut-être fixé quand j'arrêterai d'écrire des textes où j'frappe ma p'tite copine
>
> [Feminists persecute me, take me for Beelzebub / As if it's my fault that chicks are whores / They only have to stop to have their ass destroyed / And say thank you because I educate them, I teach them real stuff / Sometimes I can't remember if I'm misogynistic or if it's ironic / I'll maybe be fixed when I stop writing texts where I hit my girlfriend]

He was convicted in the first instance. The 17th chamber's judges explicitly rejected rap as a key criterion: "If verbal violence and the deliberately provocative nature of language, inherent in the musical genre that is rap, must be taken into account, [. . .] it should be noted [. . .] that freedom of art cannot be used as an absolute excuse to allow anyone who avails himself of this principle to

---

cultivant et s'appropriant l'insulte, la grossièreté et la violence du mot en les faisant entrer dans la rime, ne fut et ne reste avant tout qu'un mode d'expression utilisé par l'auteur de la chanson et ses interprètes pour à la fois exprimer la désolation et le mal de vivre des jeunes en banlieue, leur refus de se résigner face à des situations vécues comme injustes et perçues comme un rejet et formuler leur aspiration à un mode de vie autre que celui qui les exclut et les marginalise. »

avoid his duties and responsibilities, only because he considers himself an artist."[6] They consider that the analysis of utterances themselves should come first and give more significant results.

The rapper appealed the judgment in 2011. The Versailles Court of Appeal overturned the decision by formulating the following principle: the "regime of enhanced freedom" of art "must consider the style of artistic creation in question."[7] In this case, it refers to the "style" of rap, which in its view must be a crucial criterion to analyse how provocative the statements are, because it induces an inherent *distanciation*. In one case, the judges decided that the meaning of the words should be weighted more heavily than the context, and it is the exact opposite in the second case, even though they examined the same words. Other aspects were considered in those two cases, but their justification by the principle of inherent rap *distanciation* is far too imprecise and questionable, because, depending on the courts, they do not apply the same hierarchy in the criteria they use. Furthermore, this inherent rap *distanciation* is denied in some other recent cases. We can consider the examples of rappers Jo le Phéno, prosecuted for inciting against the police, and Nick Conrad, prosecuted for inciting against white people.

3.1.3 The case of Jo le Phéno

Jo le Phéno was prosecuted in 2017 for criminal incitement against the police and convicted.[8] The contentious lyrics and images are found in the music video "bavure" ["Police Blunder."] Here are some of the lyrics: "Où sont les Condés, on va les taper" ["Where are the cops, we will hit them"], "les cités sont déchainées / Et vont se défouler sur la flicaille / Elles veulent pas boire du lait au petit déjeuner / Mais juste se farcir de la volaille" ["The suburbs are raging / and will vent their anger on cops. They don't want to drink milk, they just want to stuff the poultry" ("stuff the poultry" is a play on words, because in French

---

6 TGI Paris, ch. 17, 31 mai 2013, Ministère public c/ Cotentin. In French: « Si la violence verbale et le caractère délibérément provocateur du langage, inhérents au genre musical qu'est le rap, doivent être pris en compte, [. . .] il convient cependant de rappeler que la liberté de création artistique ne saurait valoir excuse absolutoire et permettre à quiconque s'en prévaut d'échapper, au seul motif qu'il se considère comme un artiste, aux devoirs et responsabilités qui lui incombent. »
7 CA Versailles, ch. 8, 18 février 2016, RG 15/02687, Cotentin c/ Chiennes de garde et autres.
8 TGI Paris, ch. 17, 8 décembre 2017, Francis Tremblay, Agent judiciaire de l'État, Syndicat des cadres de la sécurité intérieure c/ Jois Ndjibu Milambu (dit Jo le Phéno).

cops are called "chickens")], "ce qu'on veut c'est la guerre pas le combat" ["what we want is war not only a fight"]. The meaning of these utterances is clear and can be interpreted as criminal incitement against the police. But, if we regard them within the context of the rap genre, we find the same elements as in the Sniper judgment: the expression of the feeling and claims of a social group who has had enough of what they consider police injustices against that same social group (suburban youth) and the constant use of violent and raw words. That is exactly what le Phéno's attorney pleaded, as we can read in the judgment: "He argued, in essence, that the content pursued was part of a musical genre, rap, which claimed to be a discourse of identity and protest, and which was above all perceived as a violent genre; that in this case the theme addressed by the disputed video, i.e. police violence, constituted an essential social issue on which Jo le Phéno had sought, through the use of artistic creation, to raise collective awareness." The wording is almost the same as in the considerations we read above. But in le Phéno's case, the court did not accept the argument regarding the specificity of rap language and reception. It repeatedly pointed out that he refused it. The judges answered that le Phéno "cannot, therefore, as he did at the hearing, hide behind the fact that people who know rap and its codes could not, in his opinion, interpret his music video as a call for violence against the police." Further in the judgement, they underline the fact that for them (I quote and translate) "the claim of an artistic genre, even if it is commonly accepted, like rap [. . .] as clearly identity-based, protesting and violent, cannot therefore be recognized on its own as a cause for exoneration from all responsibility." The judges concluded in this case that there is no *distanciation* "on its own." That means that according to them, the rap genre does not include an inherent *distanciation* and that it is not a key criterion. According to them, unlike the appeal courts of Rouen and Versailles, the genre is not a sufficient criterion to determine the interpretation of the utterances. It means that the methodological framework to analyse the utterances is not always the same and not always based on the same linguistic criteria. In this case they are more based on the wording itself than on the context.

3.1.4 The case of Nick Conrad

Nick Conrad was prosecuted in 2018 for criminal incitement against French white people and convicted,[9] following the online broadcast of a music video

---

9 TGI Paris, ch. 17, 19 mars 2019, AGRIF, LICRA c/ Conrad Moukouri Manga Moussole (dit Nick Conrad).

called PLB "Pendez les blancs" ["Hang white people"]. Here is an excerpt from the contentious lyrics: "Je rentre dans les crèches, je tue des bébés blancs, [. . .] pendez leurs parents, écartez-les pour passer le temps. Fouettez-les fort, faites-le franchement. Que ça pue la mort, que ça pisse le sang." ["I go into the nurseries, I kill white babies, [. . .] hang their parents, quarter them to pass the time. Whip them hard, do it wholeheartedly. So it stinks of death, so it bleeds all over"]. These utterances taken by themselves incite to commit a crime against white people. Nick Conrad argued similarly to his predecessors Orselan and Sniper, saying that the rap genre must enjoy special protections because the genre characteristically includes verbal abuse. I quote and translate the defendant's argument: "The genre of rap can require violent and brutal expression, in order to denounce the injustices and discriminations suffered by certain categories of the population. So a rap song must be analysed by taking into account the creator's subjective intention and the excessive language that this genre allows." In addition, Nick Conrad's attorney gave other criteria of *distanciation*, provided by the fact that the rapper makes cultural references to Afro-American culture (in naming figures such as Malcom X) and that he aims to denounce racism against his community. But in Conrad's case the court ruled that violence was too obvious and that the criteria of genre were not enough to consider it as a *distanciation*.

These differences in the scope of *distanciation* and in the way it is considered inherent to the rap genre introduce legal uncertainty in how creative discourses are judged by courts. The social challenge of defining accurate criteria of *distanciation* is a crucial point, especially at a time when some minorities are constantly denouncing their own unequal treatment. For the moment, it appears that the measure of *distanciation* rationalizes the preferred finding, rather than determining the finding. These examples of the way *distanciation* in rap is not considered in the same way in other cases leads us to another important question, for which I currently have no answer. By identifying recurrent protecting categories, we highlight the fact that jurisprudence is not constant. It leads us to measure the imbalance in these proceedings between racial or social characteristics of the defendant. In addition, it leads us to ask whether the sentences change according to the status of the targets. Is targeting women more permissible than targeting white people, for example? One of the aims of a quantitative analysis of legal proceedings in France is to answer such urgent questions.

4 Conclusion

This paper has focused on the rap music genre, but exactly the same issues emerge there as with other categories used by courts to measure a *distanciation*. Jurisprudence has thus forged categories of genres or registers, such as *satire*, *humour* or *fiction*, which lead to *distanciation*, even if they are not applied consistently from one case to another. It appears that the measure of *distanciation* currently serves to rationalize the preferred finding, rather that helping determine the finding. As such, these categories pose significant definitional and delimitational problems. In brief, we can point out the fact that the comparison between legal proceedings against artworks gives a worrying result. Courts do not have the same criteria to enforce freedom of art because their categories are not clarified. Therefore, this legal uncertainty cannot be solved without a dialogue between law and linguistics, which could help to unify the courts' methods. But unlike in Northern America, there is no linguistic expertise in French courts. It is very hard to convince jurists that they could need support from linguists in some cases. Still, I think that discourse analysts, literature theorists and lawyers should work together in order to make the legal framework of freedom of art clear and accurate. Working on a corpus of legal proceedings in freedom of art shows that it is urgent that literary and linguistic experts on one side, and lawyers on the other side, move forward hand in hand. For discourse analysts, such a dialogue would make it possible to compare and refine theoretical hypotheses in the light of testimonies from real readings, while lawyers could find more rigorous categories of analysis in the work of academics.

References

Achard-Bayle, Guy & Marie-Anne Paveau. 2008. Présentation. La linguistique 'hors du temple.' *Pratiques* 139–140. 3–16. https://journals.openedition.org/pratiques/1171. (accessed 05 February 2021).

Artous-Bouvet, Guillaume. 2012. *L'exception littéraire*. Paris: Belin.

Arzoumanov, Anna. 2021. *La Création en procès*. Paris: Garnier. In press.

Arzoumanov, Anna. Les catégories de l'identification et de la distanciation dans les procès de fictions. In Anna Arzoumanov, Arnaud Latil & Judith Sarfati-Lanter (eds.), *Le démon de la catégorie*, 197–210. Paris: Mare et Martin.

Arzoumanov, Anna. 2018. Les frontières du discours indirect libre au tribunal: Aperçu de la jurisprudence contemporaine en droit de la presse. *Fabula*, rubrique "Colloques en ligne," "Marges de l'indirect libre." http://www.fabula.org/colloques/document5412.php (accessed 05 February 2021).

Arzoumanov, Anna & Arnaud Latil. 2017. Juger la provocation onirique: Éléments pour une interprétation des expressions polyphoniques. *Juris Art etc* 50. 38–43.
Barbéris, Isabelle. 2019. *L'art du politiquement correct*. Paris: PUF.
Bigot, Christophe. 2017. *Pratique du droit de la presse*. Paris: Victoires Edit.
CA Rouen, ch. corr., 14 déc. 2005, procureur de la République près TGI Rouen c/ A. Karl Junior et consorts.
CA Versailles, ch. 8, 18 février 2016, RG 15/02687, Cotentin c/ Chiennes de garde et autres.
Droin, Nathalie. 2016. Le juge et le rap. *Revue du droit public et de la science politique en France et à l'étranger* 5. 1377f.
Droin, Nathalie. 2019. Rap et débat d'intérêt général: Quand le juge mêle utilement rigueur et bienveillance. *Légipresse* 374. 475f.
Edelman, Bernard & Nathalie Heinich. 2002. *L'art en conflits*. Paris: La Découverte.
Freemuse. 2020. Security, creativity, tolerance and their co-existence: The new agenda of European freedom of artistic expression. *Freemuse*. 10 January 2020. https://freemuse.org/news/the-security-creativity-tolerance-and-their-co-existence-the-new-european-agenda-on-freedom-of-artistic-expression/ (accessed 05 February 2021).
Heinich, Nathalie. 2014. *Le paradigme de l'art contemporain*. Paris: Gallimard.
Hochmann, Thomas. 2013. L'interprétation juridictionnelle du texte littéraire. In Christine Baron (ed.), *Transgression, littérature et droit*, 23–34. Rennes: Presses universitaires de Rennes.
Husson, Anne-Charlotte. 2015. "Mot-écran" et linguistique folk (1/2). *(Dis)cursives* [Carnet de recherche] 26/05/2015. https://cursives.hypotheses.org/117 (accessed 05 February 2021).
Latil, Arnaud. 2014. *Création et droits fondamentaux*. Issy-les-Moulineaux: LGDJ.
Niedzielski, Nancy & Dennis Preston. 2003 [2010]. *Folk Linguistics*. Berlin: Mouton De Gruyter.
Observatoire de la liberté de creation. https://www.ldh-france.org/sujet/observatoire-de-la-liberte-de-creation/ (accessed 05 February 2021).
Paveau, Marie-Anne. 2008. Les non-linguistes font-ils de la linguistique? *Pratiques* 139–140. http://pratiques.revues.org/1200 (accessed 05 February 2021).
Pierrat, Emmanuel. 2018. *Nouvelles morales, nouvelles censures*. Paris: Gallimard.
Ramond, Denis. 2011. Liberté d'expression: De quoi parle-t-on? *Raisons Politiques* 44. 97–116.
Reid, James H. 1998. Socializing the autonomous work of art: Pierre Bourdieu's *Les règles de l'art*. *French Forum* 23 (3). 353–370.
Sapiro, Gisèle. 2019. Repenser le concept d'autonomie pour la sociologie des biens symboliques. *Bien symboliques* 4. 2–50. https://revue.biens-symboliques.net/327 (accessed 05 February 2021).
Talon-Hugon, Carole. 2019. *L'art sous contrôle*. Paris: PUF.
TGI Paris, ch. 17, 31 mai 2013, Ministère public c/ Cotentin.
Treppoz, Édouard. 2011. Pour une attention particulière du droit à la création: L'exemple des fictions littéraires. *Recueil Dalloz* 36. 2487–2494.
Treppoz, Édouard. 2015. Photographie: La liberté de création, nouvelle limite au droit d'auteur? *Juris art etc*. 26. 6.
Tricoire, Agnès. 2003. La censure en toute légalité. *Ligue des droits de l'Homme* 31.01.2003. http://www.ldh-france.org/La-censure-en-toute-legalite/ (accessed 05 February 2021).
Tricoire, Agnès. 2015. Liberté de création: quelles menaces? quelles avancées? Entretien avec Agnès Tricoire, Propos recueillis par Lisa Pignot. *L'Observatoire* 46. 3–9.
UNESCO. Artistic freedom. https://en.unesco.org/creativity/sites/creativity/files/artistic_freedom_pdf_web.pdf (accessed 05 February 2021).

Victoria Guillén Nieto
"What else can you do to pass . . . ?": A pragmatics-based approach to *quid pro quo* sexual harassment

1 Introduction

Sexual harassment is a form of emotional abuse: victims experience high rates of psychological distress, disruption in work life and physical illness (Keashly 2001). Under the umbrella concept of sexual harassment, theorists have distinguished three related categories of sex-based harassing behaviour: gender harassment, unwanted sexual attention, and sexual coercion (Galdi, Maass, & Cadinu 2014). The category of gender harassment refers to acts aimed at creating an intimidating or hostile environment. The category of unwanted sexual attention refers to unwelcome, offensive acts of sexual interest such as objectifying gaze, unwanted touching, pressure for romantic dates, among others. The category of sexual coercion refers to predatory sexual behaviour. Typically, the perpetrator exercises power abusively over a subordinate to engage in sexual activity. This form of harassment includes behaviours such as sexual blackmail, extortion, bribes, threats, and promises of rewards aimed at inducing a sexual favour. *Quid pro quo* sexual harassment falls into the category of sexual coercion, because it refers to situations in the workplace in which individuals who hold power offer, or merely hint, that they will provide advantages or withhold disadvantages in return for the target's satisfying of a sexual demand (Scarduzio & Geist-Martin 2010).

Dealing with sexual harassment is not an easy task for various reasons. At times, neither aggressors nor victims, especially in patriarchal societies driven by sex-based ideologies of discrimination, can recognise and label incidents as sexual harassment (Dick 2013; Nava 2018). Other times, the victim of sexual harassment is uncertain about the aggressor's malicious intent. It may be also the case that the victim will be reluctant to report it to the hierarchy at work or to file a complaint, due to embarrassment, fear of not being believed, retaliation, or from lack of trust in the organisation (Shor 1992; Baron & Neuman 1998; Chamberlain, Crowley, Tope & Hodson 2008; Lutgen-Sandvik & Tracy 2012; Dick 2013). When the victim finally does dare to report, proving an episode of sexual harassment before administrators or a court of justice is

---

Victoria Guillén Nieto, University of Alicante

difficult because, unlike felony sexual assault, sexual harassment leaves no ostensive trace in the victim.

This paper aims at demonstrating the value of a micro-language approach to the legal analysis of sexual harassment. The fact that sexual harassment has more often than not a covert nature poses a challenge to the victim, who must convince administrators, or a court of justice, that the perpetrator's abusive behaviour was so pervasive that reasonable people would have perceived it as creating an intimidating or hostile environment, capable of causing harm to the dignity of the target (Scarduzio & Geist-Martin 2010; Dick 2013). The aggressors' malicious intent is likely to be hidden under the protective mantle of ambiguity and indirectness. Significantly, the indirectness in itself demonstrates the speaker's awareness that their intent is unlawful and stigmatised. As a result of the covert nature of language, the victims are also likely to doubt their own inferences as to actual intent (Baron & Neuman 1998).

The role of forensic linguists is to assist triers of fact in understanding the linguistic evidence in a case before they make a final judgement. Therefore, we hypothesise that a pragmatics-based method can improve understanding of the covert nature of *quid pro quo* sexual harassment. At the heart of the discussion is the aggressor's intended meaning. The paper aims at answering two questions: Which language clues may provide evidence of *quid pro quo* sexual harassment? To what extent may administrators or a court of justice admit such language-clues as factual evidence? The discussion begins by delineating the legal definition of sexual harassment in US and European jurisdictions. Second, we provide a literature review of related work in the field of sexual harassment. Third, we propose a pragmatics-based method for the analysis of cases involving sexual harassment. A paradigmatic case involving campus sexual harassment will serve as a basis for verifying the efficacy of a pragmatics-based method in disclosing the perpetrator's *illocutionary* act and its intended *perlocutionary* effects.

2 The legal concept of sexual harassment

Sexual harassment is a widespread but generally underestimated type of gender-based discrimination. It is, in effect, typified as a minor crime (misdemeanour) in the laws of modern societies all over the world, despite the severe consequences it may have for the target. More specifically, the US Equal Employment Opportunity Commission Guidelines define sexual harassment as "unwelcome sexual conduct when submission to such conduct is made either explicitly or implicitly a term or condition of an individual's employment."[1] Therefore, one may argue that sexual harassment is a type of sex-based discrimination that violates Title VII of the Civil Rights Act of 1964,[2] which affords employees the right to work in an environment free from discriminatory intimidation, ridicule, and insult on the grounds of sex, religion, or national origin. The US Equal Employment Opportunity Guidelines distinguish between two types of sexual harassment: *quid pro quo* and *hostile environment*. Whereas the former implies sexual coercion, the latter is related to gender harassment and unwanted sexual attention.

On the other hand, EU laws seem to have a more gender-oriented view of the problem of sexual harassment than US laws do. The Council of Europe Convention on Preventing and Combating Violence against Women and Domestic Violence (2011) typifies sexual harassment as a form of gender-based violence that constitutes a serious violation of the human rights of women, and a breach of the principle of equal treatment of men and women in the workplace.

3 Related work on sexual harassment

Research on sexual harassment is well documented in the fields of philosophy and law. For example, philosopher Vrinda Dalmiya investigated the principles that justify a moral condemnation of sexual harassment. Specifically, the author analysed the ethical dynamics of sexual harassment under three models: The *barter model*, the *boxed-in model* and the *annihilation model*. According to Dalmiya (2002), the barter model explains that sexual harassment is a matter of extorting a sexual price in exchange for some professional return. The immorality of sexual barter lies not in the sexual price that is demanded, but in the fact that the perpetrator imposes it on the target, thereby depriving that person of her free will. Under the boxed-in model, the immoral act results from the use of coercion – that is, the use of verbal, physical or moral force to compel a person to do something against her will, thus neglecting the person's dignity and sexual autonomy. Lastly, the annihilation model shows that sexual harassment reduces the target to a sexual stereotype that emerges from specific power structures. According to Dalmiya, these power-structures "perpetuate male dominance over women" (2002: 57). In the same line of thought, but adopting a legal perspective, Wright (2008) researched the legal and ethical codes that are relevant to

---

[1] https://www.eeoc.gov/laws/guidance/policy-guidance-current-issues-sexual-harassment (accessed 12 October 2020).
[2] https://www.eeoc.gov/statutes/title-vii-civil-rights-act-1964 (accessed 12 October 2020).

sexual harassment and concluded that this infringes the fundamental rights of the victim. Similarly, upon analysing the legal provisions that tackle the social menace of sexual harassment of working women, Pandey (2009) drew attention to the fact that sexual harassment is a form of discrimination on the grounds of sex, and a demonstration of power on the part of the aggressor aimed at subordinating the target, thereby violating her dignity.

Philosophy and law have made substantial contributions to the study of sexual harassment. Nevertheless, the most significant body of literature on sexual harassment comes from organisational communications research. Sociologists have contributed significantly to an improved understanding of the ideological reasons that may explain the social phenomenon of sexual harassment, the social and organisational practices that enhance such behaviours in the workplace, the surface manifestations of sexual harassment, and their effects on victims. Most sociologists have adopted macro and meso perspectives on the phenomenon of sexual harassment. To illustrate this point, let us take, for instance, the research of Chamberlain, Crowley, Tope and Hodson (2008) who investigated the intersectionality between the variables of "work power," "work culture" and "gender composition." In their view, whereas patriarchy and gender socialisation are useful concepts to explain the cultural foundations underlying sexual harassment and the victims' responses, the organisational context governs whether, and how, sexual harassment occurs in a given workplace. Scarduzio and Geist-Martin (2010), for their part, analysed the influence of ideological positioning on four male professors' accounts of their experience with sexual harassment. Their research drew attention to the fact that, when investigating sexual harassment, it is necessary to understand the "gendered" nature of power concerning sexual harassment and ideological positioning.

However, in our view, the most ambitious research is that of Lutgen-Sandvik and Tracy (2012) who adopted a multi-layered approach. That is, they adopted an approach involving macro, meso and micro levels of analysis. In their view, sexual harassment in the workplace "is condoned through societal discourses, sustained by receptive workplace cultures, and perpetrated through local interactions" (2012: 3). Their in-depth analysis provided answers to several essential questions, such as: how emotional abuse manifests itself in the workplace, how employees respond, why it is so harmful, why the resolution is so difficult, and how administrators might resolve it. Although it is relatively uncommon in the field of organisational communication research, some theorists have adopted a micro perspective when analysing the target's experiences. Such is the case of Keashly, who directed attention to the fact that the crime often becomes "unidentifiable and likely not punishable" (2001: 240) because of its ambiguous nature.

Other theorists in organisational communication research have provided compelling insights on sexual harassment from a psychological perspective.

For example, Malovich and Stake (1990) analysed the personality traits of aggressors and victims of sexual harassment in the workplace. Their research tested the relationship between sexual harassment attitudes and the personality variables of self-esteem and sex-role attitudes. The authors mentioned concluded that personality factors are essential in determining harassment attitudes, and in identifying individuals prone to harass or tolerate harassment. Dick (2013) focussed on the way sexism affects the individual's discursive practices and their understanding of sexist behaviours such as sexual harassment. According to Dick, sexual harassment has a dualistic nature, because it can be understood simultaneously as both a subjective interpretation and an objective reality. On the one hand, the subjective nature of sexual harassment is evident in the fact that the burden of the proof lies on the victims who must report on behaviour that they perceive as offensive. On the other hand, the objective nature of sexual harassment also operates as a self-regulatory mechanism because it is socially stigmatised. There are also studies explaining the role played by the arts in romanticising patriarchal values and unequal power of men and women (Shor 1992), or showing the way media content plays a central role in activating harassment-related social norms, which in turn encourage or inhibit harassing conduct (Galdi, Maass & Cadinu 2014).

On reviewing the literature, we find that while research on sexual harassment principally comes from the social sciences, the crime, which is mainly perpetrated through malicious language use, has not yet received sufficient attention as a scientific object of study in linguistics. *The language of sexual crime* (Cotterill 2007) and *The language of sexual misconduct cases* (Shuy 2012) are among the few linguistic studies relating to the topic. Whereas the former is mainly devoted to the analysis of the language of consent in rape law, the latter concentrates on the linguistic analysis of sexual misconduct cases in the workplace. Shuy addresses several prosecutorial problems in gathering and analysing evidence – e.g. inadequate electronic recordings, inadequate transcriptions of recordings, inadequate police interviewing, and misinterpretation of the actions and intentions of both the complainants and suspects, among others. Shuy proposes a top-down language-based method of analysis utilising the following linguistic tools: (a) speech event analysis, (b) schema analysis, (c) agenda analysis as revealed by topics and responses, (d) speech act analysis, (e) cooperative principle analysis, (f) conversational strategies analysis, (g) recency principle analysis, (h) analysis of ambiguity, specificity and referencing, and (i) analysis of grammatical and lexical features.

4 Method

Since analysing the perpetrators' intended meaning in the context of utterance is at the heart of cases involving sexual harassment, we can reasonably argue that pragmatics, the branch of linguistics that studies how context contributes to meaning, is optimal for the analysis of this type of language crime. We propose a top-down pattern of analysis. First, we identify the *speech event* represented by the language evidence. Second, we identify deviations from the *mental frame*, or *schema*, people have of the speech event. Third, within the framework of (im)politeness theory, we determine whether the deviations observed may be indicative of offensive behaviour (*face attack*) on the part of the suspect that causes *face loss* to the target. Finally, we turn to speech act theory (Austin 1962; Searle 1969; 1979; 1983) and Grice's *cooperative principle* (1975) in order to interpret the suspect's intended meaning (*illocutionary act*) in the context of utterance.

The pragmatics-based method we propose is illustrated through its application to an exemplary case of campus sexual harassment. We wanted to test the definition of harassment as behaviour that "reasonable people" would consider harassment, so we conducted a small survey. This method involved collecting data from a pre-defined group of respondents. The group consisted of 65 Spanish students taking the last year of a university degree. Their ages ranged between 21 and 25 years old. While 53 were female students, 12 were male students. They all volunteered to take part in the survey. The survey's purpose was to analyse their expectations, emotional reactions, and their interpretation of a speaker's intended meaning in a hypothetical communicative situation. We asked the students to answer a questionnaire consisting of seven items:

Item 1 asked the students to tick off from a list of actions the ones that they would not expect in the speech event of "an exam."

Item 2 enquired about the students' emotional response in a hypothetical situation: "You are about to do an exam in a professor's office and all of a sudden, he locks the door."

Item 3 enquired about the students' emotional response in a hypothetical situation: "While you are doing an exam in a professor's office, the professor tells you that you are going to fail."

Item 4 enquired about the different solutions the students would offer to make it possible for them to pass the exam, in the situation described in item 3.

Items 5 to 7 enquired about other alternative solutions the students can think of in order to pass the exam in the situation described in item 3.

5 Revisiting a case of campus sexual harassment

At the end of the academic year 2015–16, a Chinese female postgraduate student (henceforth referred to as *the target*) filed a complaint before a Rectorate in a Spanish University.³ The complaint was against a male professor (henceforth referred to as *the suspect*) who had allegedly sexually harassed her while she was writing an exam in his office. The student appended to her complaint a 20-minute audio recording she had managed to make of the incident. The administrators admitted the complaint, in compliance with Spanish Organic Law for Effective Equality of Women and Men (B.O.E. Nº 7123/03/2007), and activated the protocol against campus sexual harassment. The Rectorate established a Joint Consultative Committee for purposes of determining whether, or not, the target had been a victim of sexual harassment by the suspect. The Committee was composed of six members: three male professors and three female professors. The result of the investigation was inconclusive. While the three male professors concluded that the complainant had not been a victim of sexual harassment, the three female professors concluded that she had been. The clash of opinions between the male and female professors evaluating the case illustrates the difficulty involved in perceiving sexual harassment objectively, and illustrates also the political process through which particular versions of reality acquire authority. This political process is, according to Dick, "critical for understanding the reproduction, resilience and endurance of social facts such as sexism" (2013: 645).

Two years later, in an attempt to find a reasonable solution to the case, and thereby close the administrative procedure, the Rectorate decided to hire the professional service of two forensic experts. First, the administrators asked an engineer to authenticate the audio recording provided by the complainant. After analysing the audio recording, the expert certified that it had not been subject to manipulation. Second, the administrators asked a linguist to determine whether there was linguistic evidence to support the claim that the context had been intimidatory, degrading and offensive for the complainant and, at the same time, whether there was linguistic evidence to support the hypothesis that the complainant may have been a victim of sexual coercion. The administrators gave the expert linguist the following materials: (a) an authenticated audio file (mp3) (7.263.571 bytes), whose duration was 20 minutes approximately, and (b) the transcript of the audio recording elaborated by the Secretary of the Joint Consultative Committee.

The linguist found that the transcript was inadequate because, although it contained the conversation held between the subjects, it failed to provide any

---

3 For data protection issues, all personal references have been eliminated in the paper.

details about the way they had expressed their utterances. As a result, the linguist had to rewrite the transcript of the audio recording with the assistance of a computer system – Praat: Doing Phonetics by Computer v 6.1.0.2 (Boersma & Weenink) – and the international transcription symbols for Conversation Analysis. As a result, apart from the content of the audio recorded conversation, the improved transcript included exhaustive information about speech prosodic and para-linguistic features that are key elements in determining subjects' emotions, attitudes, and intended meanings in social interaction. Upon listening to the audio recording, we learn that the conversation starts in medias res, specifically at the point the suspect is asking the target "to offer him a solution." The content of the audio recording revolves around the discourse topic of "making an offer to pass the exam." The conversation structure takes the shape of a loop because the suspect recurrently goes back to the initial question. The linguist can listen to the professor's voice conveying self-control and, to some extent, passive aggression, and the target quietly sobbing. As announced earlier, at the heart of this discussion is the suspect's intended meaning when he asks the student questions such as: "Why don't you give me a solution?"; "What can you do to pass?"; "What else can you do to pass?"; "Another thing?"; "You cannot do anything else?"; "I can fill in the test for you, but then what do I get in exchange?"; "What do I get in exchange if I pass you?"[4]

6 Analysis and discussion

6.1 Identifying the speech event

The speech event refers to the basic unit of interactive verbal behaviour that is socio-culturally driven by unspoken conventions or inner rules (Gumperz 1990). The speech event is bound to a speech situation or social context of interaction, which includes, among others, the following typified categories: (a) the physical setting, (b) the social context or scene, (c) the purpose, (d) the act sequence, (e) the mode of communication – that is to say, the channel, the register, and the tone, (f) the social norms, and (g) the genre (Hymes 1974). Based on these typified categories, one can predict where the speech event may, or may not, take place; who may, or may not, participate in the speech event; what communicative purposes the participants may, or may not, have; which language may,

---

[4] This is an English translation of the questions the professor asked the student in Peninsular Spanish.

or may not, be permissible; what topics may, or may not, be spoken about; and what may, or may not, happen, among other aspects. One may therefore argue that the speech event activates a mental frame, or schema, based on people's lived experience, that sets out "structured expectations" (Tannen 1993) about such a speech event. People use these structured expectations as guidance for understanding and interpreting meaning in any communicative situation.

Upon analysing the audio recording provided by the target, the speech event "an exam" emerges. As soon as people mention such a speech event, a schema is activated in their minds that they call upon as they try to understand and interpret new information that comes to them. A schema works as a baseline – that is, as a fixed point of reference that people use for comparison purposes, and serves to distinguish a particular speech event from other speech events typical of academe, such as lectures, seminars, workshops, interviews, among others. We can then predict that "an exam" as a speech event is bound to a speech situation, including the following typified contextual elements (Hymes 1974):

a) Physical setting. The exam may take place, for instance, in a classroom or in a meeting room. The exam may be held either in the morning or in the afternoon, on a school day.
b) Social context or scene. People expect that the participants in the speech event "an exam" take up the social roles of professors and students. People also expect them to have an asymmetrical power relationship. In other words, while professors have institutionally protected power to ask questions to students, and to assess their work, students do not have the power to do so, but, instead, must try to answer the questions that their professors ask.
c) Purpose. Whereas the purpose of students is to get a mark that is mandatory for their graduation, the purpose of the professor is to assess students' knowledge and abilities in a particular subject.
d) Act sequence. From the lived experience of the researcher, a typical act sequence may include acts such as: "the professor and students greet each other," "the professor distributes the exams to the students," "the professor gives the students instructions about the exam," "while the students answer the exam questions, the professor watches in silence the progress of the exam," among others.
e) Mode of communication. The mode of communication may be oral or written. In either case, one expects the participants to use a formal register and a respectful tone.
f) Standards of conduct. One expects professors to watch the exam, in order to guarantee that all students have the same opportunities to answer the questions. The students are aware of being under surveillance by the

professor. Both professors and students must be civil and respectful to each other.
g) Genre. One expects the participants to use academic genres as instruments of social communication in the speech event – e.g. "an exam," "giving instructions," "an opinion essay," "a report," "a summary," among other possibilities.

The recognition of the typified contextual elements of the speech event acts positively on the subjects who interact in a given communicative situation, because it reinforces their expectations and thereby grants them certainty and security.

6.2 Identifying deviations from the mental frame of the speech event

While the recognition of the common elements of a speech event acts positively on the subjects who interact in a communicative situation, the presence of unexpected elements is likely to break the subjects' expectations, causing them disorientation and uncertainty. Deviations from the baseline may be due to socio-cultural variation, or they may be intentional. Upon analysing the audio recording, one can identify significant deviations from the schema of "an exam." These deviations affect the following contextual elements:

a) Physical setting. The exam is taking place in an office. At the end of the audio recording, we learn that the suspect has locked the office door, so that it is not possible for the target to leave the office unless the suspect opens the door for her.
b) Social context or scene. As already mentioned, the participants in the speech event are a male professor (the suspect) and a female Chinese postgraduate student (the target). The audio recording points to a power asymmetry between the subjects that exceeds what is considered socially acceptable.
c) Purpose. As expected, the purpose of the target is to sit for an exam. However, the purpose of the suspect seems to be somewhat ambiguous, because he introduces the element of solicitation. Specifically, the suspect is pervasively trying to obtain something from the student in return for a pass.
d) Act sequence. The act sequence is not linear, but instead revolves around the act of solicitation (Keashly 2001).
e) Key. The suspect uses a patronising tone with the target. He calls her *hija* 'darling,' and treats her as if she were a little girl, rather than as a postgraduate student.

f) Standards of conduct. It is noteworthy that the suspect violates the standards of conduct (Keashly 2001) when he shows no respect for the target. Specifically, he humiliates the target by telling her she is incapable of passing without his help. Moreover, the suspect interrupts the target abruptly when she tries to reply to his questions, and he even shouts at her.
g) Genre. Unexpectedly, the suspect introduces the genre of "negotiation," specifically the phase of bidding and bargaining.

In order to verify what reasonable people would perceive as unexpected under similar circumstances, item 1 of the questionnaire asked the students surveyed to tick off the actions on a list that they would not expect in this situation: "Imagine that due to serious family health issues, you must go back home all of a sudden. The professor of the subject allows you to sit for the exam in his office before you go home." Figure 1 below shows the distribution of percentages per sex regarding the unexpected acts that were identified by the group of students surveyed.

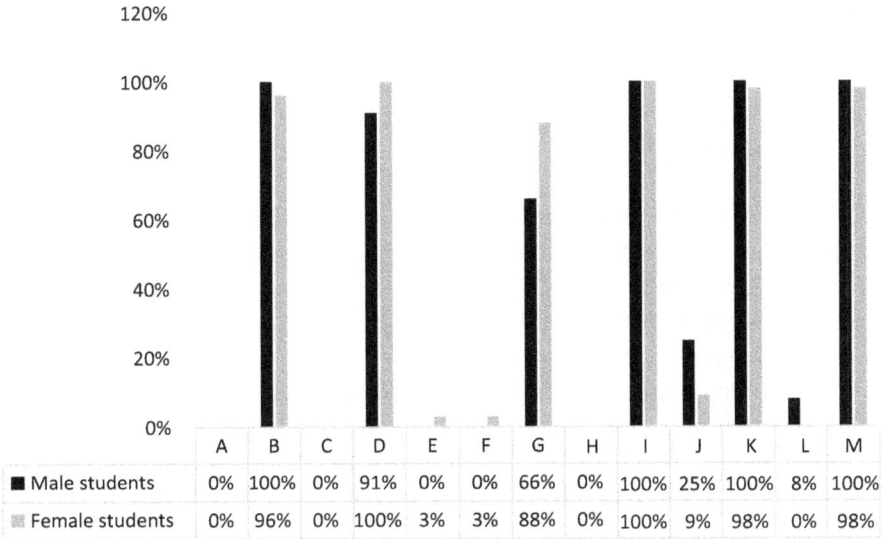

Figure 1: Unexpected acts identified by the students surveyed.

As shown in Figure 1, the students ticked off the following actions as unexpected in the communicative situation: (B) The professor locks the door (100%–96%), (D) The professor tells me that I am going to fail (100%–91%), (G) The professor harasses me to finish the exam on the excuse that he must go (88%–66%), (I) The professor addresses me as "my darling" and treats me as if I were a little kid (100%), (K) The professor solicits from me a proposal to pass the subject (100%–98%), and (M) The professor threatens to fail me if I do not give her/him a solution (100%–98%).

Up to this point, we have identified the speech event that the language evidence represents. According to Shuy, becoming aware of and understanding the speech event should be "the first thing for prudent prosecutors and defence attorneys to do when they review the language evidence in a case" (2012: 196). Furthermore, we have observed that the suspect's purpose is to elicit from the student an offer of something in exchange for passing the exam. The suspect's purpose thereby deviates from the one expected in the speech event, which would ordinarily be "assessing the student's knowledge and abilities in a given subject." The deviation is likely to cause disorientation to the target, as she tries hard to make sense of the suspect's intended meaning. In other words, deviating from the expected communicative purpose is like breaking the rules of a game, because it triggers a new set of expectations about a different speech event. We will now move on to consider whether the deviation identified is intentional on the part of the suspect.

6.3 Identifying offensive behaviour in the communicative situation

At the core of the theory of (im)politeness[5] is the concept of face – or public self-image – that Brown and Levinson (1987 [1978]) adapted from Goffman (1967). Brown and Levinson's concept of face combines two basic wants: the desire to be unimpeded (*negative face*) and the desire to be approved of (*positive face*). As a result, participants in conversation must strive to *save face*. That is, participants must try not to threaten other interlocutors' *face wants* or their own *face needs*, otherwise they will *lose face*. The problem lies in the fact that certain kinds of acts intrinsically threaten face (*face-threatening acts*), especially those acts that by their nature run contrary to the face wants of the addressee or of the speaker. For example, "orders," "requests," and "offers" intrinsically threaten the addressee's negative-face wants, while "expressions of disapproval, criticism, contempt or ridicule" threaten the addressee's positive-face wants. Furthermore, some acts such as "interruptions," "threats," and "requests

---

5 Alba-Juez and Mackenzie (2016) provide a comprehensive review of (im)politeness theory.

for personal information" inherently threaten both negative- and positive-face wants of both speaker and addressee.

Participants in conversation may either exhibit polite linguistic behaviour by using language for face-saving purposes (Lakoff 1973; Brown & Levinson 1987 [1978]; Leech 1983; Watts 2003) or display impolite linguistic behaviour by using language for face-threatening purposes (Spencer-Oatey 2005; Bousfield 2008; Culpeper 2011). More specifically, according to Culpeper, impoliteness occurs when: "(1) the speaker communicates face-attack intentionally, or (2) the hearer perceives and/or constructs behaviour as intentionally face-threatening, or a combination of (1) and (2)" (2011: 9). Impoliteness offers the linguist a useful pattern for the analysis of offensive behaviour that consists of three steps: (1) appraisal of the communicative situation, (2) activation of impoliteness attitude schemata, and (3) activation of impoliteness related emotion schemata (Culpeper 2011). For example, let us consider the case under analysis. On appraising the communicative situation, the target is first likely to identify deviations from the baseline of the speech event "an exam." Second, the target is likely to perceive and construct the suspect's behaviour as an intentionally face-threatening act. Third, the target is likely to experience face loss due to the suspect's face-threatening behaviour. It is then realistic to think that because of the suspect's face threatening behaviour, the target was uncertain and she therefore recorded the conversation as a strategy of passive resistance to the suspect. In order to ascertain what reasonable people would feel under similar circumstances, items 2 and 3 in the questionnaire asked the group of students surveyed about their emotional responses. It is noteworthy that whereas both the male students and the female students would feel puzzled (100%), only the female students would also feel fearful (100%), if they were locked in an office with a male professor.

From the perspective of (im)politeness theory, offensive behaviour is linguistically realised through face-threatening acts – that is, acts that cause damage to an individual's face – public social image – producing face loss. *Face-threatening acts* can be verbal, prosodic – e.g. tone and inflexion, para-verbal – e.g. modifiers, differentiators and alternants, and non-verbal – e.g. facial expressions, gestures, body movements and postures (Poyatos 1984). At a minimum, there must be one of these acts associated with the utterance for it to be a face-threatening act.

The linguistic analysis of the audio recording reveals the presence of face-threatening acts in 49 of the suspect's 54 speech turns, causing damage to either the negative or the positive face of the target. In what follows, we will analyse, in further detail, the types of face-threatening acts performed by the suspect in the speech event.

a) Face-threatening acts causing damage to the negative face of the target

As defined by Brown and Levinson, negative face refers to "the want of every competent adult member that his actions be unimpeded by others" (1987 [1978]: 62). Negative face is threatened when an individual does not avoid or intend to avoid the obstruction of their interlocutor's freedom of choice and action. A negative face-threatening act can cause damage to either the speaker or the addressee, and makes one of the interlocutors submit their own will to that of the other. Example 1 illustrates some of the suspect's face-threatening acts, impeding the student's successful completion of the exam.

Example 1[6]
1 PROFESOR: *¿No hay ninguna solución?*
1 ALUMNA: *hh ((aspiración profunda, solloza)) ¿Pues entoce yo no puedo ((quebrada)) aprobarlo?* [. . .]
5 PROFESOR: *Ya, pero si te estoy diciendo que no es un problema ((ininteligible)). Problemas que me estás trasladando a mí, y digo ese problema – no lo sé hija. Quiero que me des una solución. Examen no, porque examen por por mucho examen que te ponga ((con enfado)) me lo vas a hacer igual. Yo te he dicho que cojo que cojo te cojo yo a ti-*
5 ALUMNA: hh ((*gimiendo*))
6 PROFESOR: *Déjame te cojo a ti te lo marco yo –* [. . .]
7 PROFESOR: *Entonce, bueno pues marcaló si quieres, marca lo que tú creas que está bien y ya está. Pues venga date prisa que nos vamos ahora mismo, eh márcalo ya porque nos vamos ahora mimo. Yo me tengo que ir eh ((habla muy rápido, atosigando)) así que ((ruido metálico)) vámonos.* [. . .]

[1 PROFESSOR: Isn't there any solution?
1 STUDENT: hh ((deep inhalation, sobbing)) So then I can't ((broken voice)) pass? [. . .]
5 PROFESOR: Yeah, but I'm telling you that is not a problem ((unintelligible)). You are passing the bucket, and I say that problem – I don't know my darling. I want you to give me a solution. Exam no, because no, no matter how many exams you take ((angrily)) you're gonna do the same stuff. I have told you that I take, that I take, I take you –
5 STUDENT: hh ((sobbing))
6 PROFESSOR: Let me take you, I take you, I fill it in for you. . . [. . .]
7 PROFESSOR: Then, well so fill it in if you want, fill in what you think is right and that's it. So, hurry up, 'cause we're finishing right away, umm fill it in 'cause we're finishing right away. I must be off, umm ((fast speech, harassing her)) so ((metallic noise)) let's go.]

---

6 For purposes of accuracy, we quote excerpts from the original transcript of the audio recording in Peninsular Spanish. The transcript shows the utterances as they were expressed by the participants in the communicative situation. Although part of the original meaning is inevitably lost in translation, the original transcripts are also translated into English to help the reader better understand the text.

In exchange 1, the suspect solicits from the student a solution to pass. The act of solicitation threatens the target's negative face, because it puts pressure on her to give him a solution. In exchanges 5–6, the suspect orders the student to give him a solution to pass, but he claims that this solution should be different from sitting for another exam because she will fail again. The suspect even tries to snatch the pencil away from the target to tick off the correct answer for her in the test. These face-threatening acts are performed on record, without redressive action with respect to the addressee's negative face want. That is, these acts are performed directly. As the suspect does so, he continues putting pressure on the target to give him a solution. In the next speech turn (7), the suspect performs a chain of negative face-threatening acts designed to impede the target's completion of the exam. More specifically, one can observe that the suspect urges the target, on the excuse that he needs to leave, to finish the exam quickly.

Interestingly enough, in Example 2 the suspect solicits an action from the target. That is, the suspect wants the target to tell him what to do. In doing so, the suspect ostensibly damages his own negative face because he submits himself to the target's will. However, what he is really doing is inducing the victim to do something for him. Next, the suspect offers to pass the target. The bid is not free from imposition on the target's freedom of choice and action, because the suspect demands some benefit in return. It is important to note that either rejection or acceptance of the suspect's bid puts pressure on the target. In the case she accepts the bid, she may also incur debt. In the case she rejects the suspect's bid, she risks failing the exam. From exchanges 52–54 we learn that during the time the student was writing the exam, the office door was locked. This non-verbal act constitutes a negative face-threatening act, because it impedes the target's freedom of movement and autonomy. That is, she can only get out of the suspect's office, if he first unlocks the door.

Example 2
49 PROFESOR: ((ininteligible)) *Pero ¿qué qui – quieres que haga? Tú di dime que quieres que haga para que yo te pueda aprobar* ((ininteligible)). *Entonces, ¿qué me llevo yo si te apruebo? ¿Qué me llevo yo si te apruebo? ¿A currar?* ((ininteligible)) ((ininteligible)) *Pues si no hacemos nada, si no hacemos nada, otra cosa hija mía no puedo hacer.* [. . .]
52 PROFESOR: *¿Quieres salir?*
52 ALUMNA: *Sí* ((débilmente)). *¿Qué hago?* ((débilmente)) *Adiós.*
53 PROFESOR: *¿Salir?*
53 ALUMNA: *Yo quiero salir.*
54 PROFESOR: *¿Quieres salir?*
54 ALUMNA: *Quiero.*
((ruido como de desenchufar un aparato)) ((ruido de cerrojo de puerta)) ((portazo))

[49 PROFESSOR: ((unintelligible)) But what wha' – what do you wanna me to do? Tell me what you wanna me to do for you to pass ((unintelligible)). Then what do I get if I pass you? What do I get if I pass you? More work? ((unintelligible)) ((unintelligible)) So if we don't do anything, if we don't do anything, another thing my darling I can't do. [. . .]
52 PROFESSOR: Do you wanna go out?
52 STUDENT: Yes ((faintly)). What shall I do? ((faintly)) Bye.
53 PROFESSOR: Go out?
53 STUDENT: I want to go out.
54 PROFESSOR: Do you wanna go out?
54 STUDENT: I do.
((noise as if unplugging a computer)) ((noise as if unlocking a door)) ((door slam))]

b) Face-threatening acts causing damage to the positive face of the target

As defined by Brown and Levinson, positive face is "the want of every member that his wants be desirable to at least some others" (1987 [1978]: 62). Positive face is threatened when the speaker does not care about their interlocutor's feelings, their wants, or does not want the other's wants. Example 3 illustrates some of the suspect's face-threatening acts that cause damage to the target's positive face wants.

Example 3
4 PROFESOR: *Por mucho que rellenes hija* – ((ruido como de pasar páginas))
4 ALUMNA: *Si tengo muchas, no, no expli* – [. . .]
10 PROFESOR: *No podemos hacer – ¿no hay otra, otra solución? Pero que así no se puede ir, así es que no. No puedes hacer las cosas así* ((tono paternalista, condescendiente)) *– sin saber bien el idioma y no entender bien las cosas. No puedes – y ya no solamente sino cuando veas* ((ininteligible)). *¡Hija!* ((habla rápida)).
10 ALUMNA: *No, esto sí* ((voz quebrada)) *yo – yo puedo entender.*

[4 PROFESSOR: No matter how much you fill in darling – ((sound as if turning pages))
4 STUDENT: If I have many no, no explai – [. . .]
10 PROFESSOR: We can't do – there's no other, other solution? But you can't go like that, not like that. You can't do things that way ((with a patronising tone, condescending)) – you don't know the language well, and you don't understand things properly. You can't – and not only that, but also when you see ((unintelligible)). Darling! ((fast speech)).
10 STUDENT: No, this, yes ((broken voice)) I – I can understand.]

In example 3, one can observe several face-threatening acts causing damage to the target's positive face want. In exchange 4, the suspect degrades the target as a student by implying that she will not pass. He also patronises the target calling her *hija* 'darling,' and he interrupts her abruptly. Subsequently, in exchange 10, the suspect continues damaging the target's positive face by degrading the target for her lack of linguistic competence in Peninsular Spanish, and he calls her *hija* again, as if she were a little girl rather than a postgraduate student. These face-threatening acts alert us to the presence of emotional manipulation with a dual purpose. First, it is reasonable to think that the suspect seeks to instil feelings of shame and guilt in the target. Second, the suspect aims at introducing changes in the target's behaviour, changes that will reap him a benefit.

In sum, the application of (im)politeness theory to the case under study has helped us to disclose the suspect's offensive behaviour. On the one hand, the suspect carries out face-threatening acts that damage the target's negative face as they impede her freedom of choice and action in the speech event. That is, the target cannot get out of the office because the suspect has locked the door; the target cannot fill in the exam because the suspect is harassing her and demanding "a solution." On the other hand, the suspect also performs face-threatening acts that damage the target's positive face. These face-threatening acts cause harm to her self-esteem, self-reliance, and to her dignity as a person. On taking a closer look at the face-threatening acts performed by the suspect in the speech event, coercion emerges. Coercion refers, in effect, to the use of emotional manipulation to persuade the target to do something she may not want to do in return for a pass.

6.4 The speaker's intended meaning and conversational implicatures

In an attempt to save face, perpetrators of *quid pro quo* sexual harassment will typically try to achieve their ultimate purpose through covert language. Covert language inevitably leads to the non-observance of the cooperative principle (Grice 1975): "Make your contribution such as is required, at the stage at which it occurs, by the accepted purpose or direction of the talk exchange in which you are engaged" (Grice 1975: 45). Rational speakers are expected to comply with the cooperative principle in social interaction. However, the truth is that speakers frequently intentionally "flout" one or more of the principle's four maxims: *the maxim of quantity*; *the maxim of quality*; *the maxim of relation*; *the maxim of manner*.[7] In these cases, the hearer still works on the assumption that the speaker is trying to be cooperative, and therefore searches for meaning at a contextual level. That is, the speaker draws an inference from the flout that is

---

7 Grice (1975: 45–46) provides a detailed explanation of the cooperative principle and its maxims of conversation.

known as *conversational implicature*. Conversational implicatures refer then to meaning that is nondetachable from the semantic content of the utterance, but instead is context-sensitive. Grice (1975) distinguished between what is *said* or *entailed* (what our words mean) and what is *implicated* (what we mean by our words). As mentioned at the outset, at the heart of the discussion is the suspect's intended meaning in the context of utterance. In Example 4, the blatant non-observance of the maxim of manner[8] on the suspect's part triggers the target's search for a conversational implicature.

Example 4
43 PROFESOR: *¿No se te ocurre nada? Venga, ¡dime algo que se te ocurra! Has hecho un examen. ¿Te voy a hacer otro examen? – es que vas a hacer iguá. ¿Otro trabajo? – un trabajo ya lo has hecho entonces* ((ininteligible)) –
43 ALUMNA: *Otro, otro trabajo.*
44 PROFESOR: *Trabajo otro – no puedo hacer porque esto este máster pasa por una auditoría, y si a mí me piden tus notas, me piden el trabajo, y el examen – es que yo quiero que entiendas, me piden tu trabajo y tu examen – si yo le doy el trabajo aprobado y el examen suspendido –*
44 ALUMNA: *y ahora examen con las pre preguntas más claro más más –* [. . .]
49 PROFESOR: ((ininteligible)) *Pero ¿qué qui quieres que haga? Tú di dime que quieres que haga para que yo te pueda aprobar* ((ininteligible)), *entonces ¿qué me llevo yo si te apruebo? ¿Qué me llevo yo si te apruebo? ¿A currar?* ((ininteligible)) ((ininteligible)) *Pues si no hacemos nada, si no hacemos nada, otra cosa hija mía no puedo hacer.*
49 ALUMNA: *Usted me me me diga qué tengo que ha hacer, qué puedo hacer.*
50 PROFESOR: *¿Y qué puedes hacer para aprobar? Tú dime que puedes hacer. ¿Qué se te ocurre?* ((gesto vocálico)) ((ininteligible)) *¿Qué se te ocurre puedes hacer para aprobar? Otra cosa no puedo hacer* ((ininteligible)) ((ruido de objeto)) *A ver – ¿tú qué crees que se puede hacer para aprobar?*
50 ALUMNA: *No tengo idea.*
51 PROFESOR: *¿No se te ocurre nada? ¿Nada de nada? ¿No se te ocurre nada? ¿No haces otra cosa?* ((ruido)) *Yo otra cosa no te puedo hacer, otra cosa no te puedo hacer.*

[43 PROFESSOR: You can't think of anything? Come on, think about something you can do! Shall I ask you to do another exam? You're gonna do the same. Another paper? – you've already done it then ((unintelligible)) –
43 STUDENT: Another, another paper.
44 PROFESSOR: Another paper – I can't 'cause this this is a master's that is evaluated, and if they ask me for your marks, they ask me for your paper, and the exam – I wanna you understand, they ask me to turn in your paper and your exam – if I turned in your paper passed, and the exam failed –
44 STUDENT: And now exam with clearer questions, clearer, clearer – [. . .]

---

8 "Under the category of manner, falls a supermaxim 'Be perspicuous' – and various maxims such as: (1) Avoid obscurity of expression, (2) Avoid ambiguity, (3) Be brief (avoid unnecessary prolixity), and (4) Be orderly" (Grice 1975: 46).

"What else can you do to pass ...?" — 49

49 PROFESSOR: But what wha' what do you wanna me to do? Tell me me what you wanna me to do for you to pass ((unintelligible)), then what do I get if I pass you? What do I get if I pass you? More work? ((unintelligible)) ((unintelligible)) So if we don't do anything, if we don't do anything, another thing my darling I can't do. [. . .]
49 ALUMNA: Please tell me me me what I have to do, what I can do.
50 PROFESSOR: And what can you do to pass? You tell me what you can do. What do you think? ((vocalic gesture)) ((unintelligible)) What do you think you can do to pass? Another thing I can't do ((unintelligible)) ((object noise)) Let's see – what do you think you can do to pass?
50 STUDENT: I have no idea.
51 PROFESSOR: You can't think of something? Nothing at all? You can't think of something? You don't do another thing? ((noise)) I can't do any other thing for you, any other thing I can't do for you.]

From exchanges 43–44, we can infer that the target assumes that the suspect is asking her for academic solutions to pass the course, and she therefore suggests several appropriate options such as "write an extra paper," "repeat the exam," or "do another exam with questions formulated in plain language." It is noteworthy that the academic options the target gives match those of the group of students surveyed in item 4. One can reasonably assume that, at least at this stage, both the 65 students surveyed and the target are working on the assumption that the suspect means what he says. Figure 2 shows the responses of the group of students surveyed.

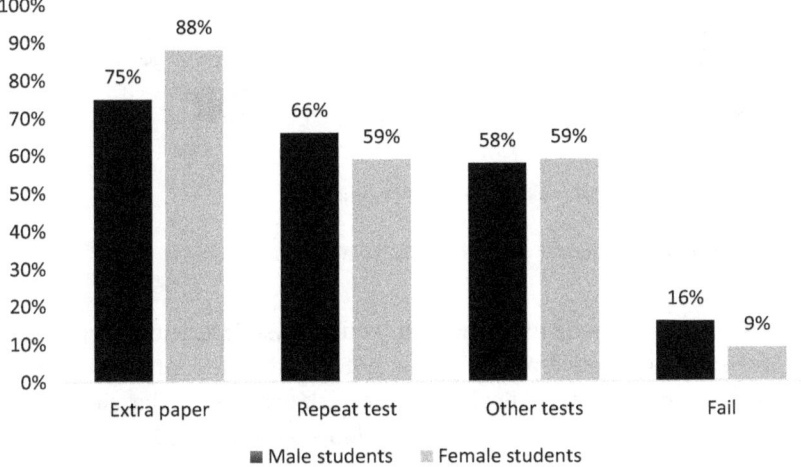

Figure 2: Options given by the group of students surveyed.

Since none of the academic options given by the target, nor by the 65 students surveyed, seems to be the preferred response the suspect wants to elicit from the target, one may infer that the suspect does not have an academic solution in mind, and therefore is not being sincere. This fact implies a non-observance of Grice's maxim of quality.[9]

When the suspect fails to achieve the desired perlocutionary effect, he continues insistently asking the target to give him a solution, as seen in exchanges 50–51. In this context, a competent speaker would assume that the suspect is trying to be cooperative, and therefore look for meaning at a deeper level than the literal one. In order to find out what reasonable people would infer under similar circumstances, we consult the answers given to items 5, 6 and 7 by the group of students surveyed. The answers of the students surveyed are shown in Figure 3.

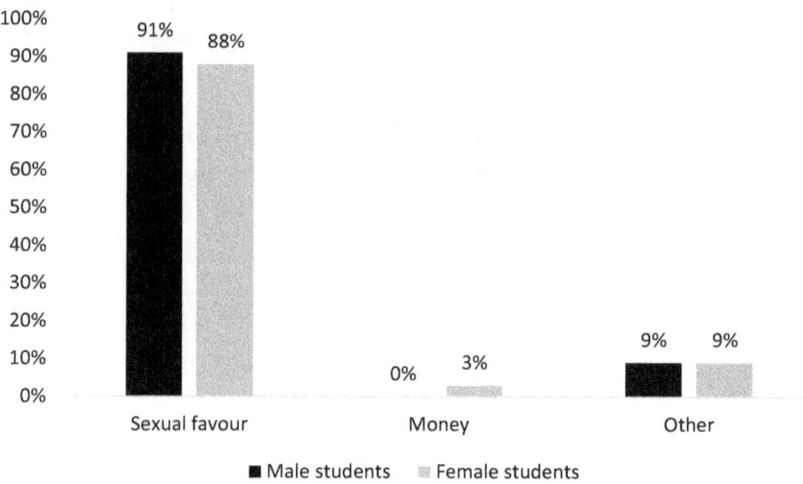

Figure 3: Students' interpretations of the utterance: "What else can you do to pass?".

Competent speakers of Peninsular Spanish would make inferences of the type called conversational implicature, which is not coded in the language but instead is context-sensitive. Specifically, Figure 3 shows that under similar circumstances, a very high percentage of the students surveyed – 91% of the male

---

9 "Under the category of quality falls a supermaxim: Try to make your contribution one that is true and two more specific maxims: (1) Do not say what you believe to be false and (2) Do not say that for which you lack adequate evidence" (Grice 1975: 46).

students and 88% of the female students – would infer that the professor in the hypothetical situation means something different from what he says. As shown from Figure 3, the vast majority of the students – 91% of the males and 88% of the females – infer that the professor is soliciting a sexual favour in return for a pass on the exam. However, it is important to note that the target – a Chinese female student with low communicative competence in Peninsular Spanish – does not seem to secure uptake of the suspect's intended meaning. Her answer to the suspect's solicitation is *no tengo idea* 'I have no clue.' We may then argue that the target's apparent failure to infer the conversational implicature poses a challenge to the property of *universability* that features in conversational implicatures. According to Levinson (2000), conversational implicatures, in contrast with coded meanings, are expected to be universal because they derive from basic considerations of rationality. The property of universability means that in any language into which a given utterance is directly translated, the equivalent form should carry the same standard implicatures. The following question then arises: Does the target not infer the suspect's intended meaning because she is a Chinese student with low communicative competence in Peninsular Spanish? However, one may reasonably infer that by saying *no tengo idea* 'I have no clue' the target shows uncooperative behaviour, as a way to fight off sexual coercion. (Indeed, we should not forget that, in resistance to the suspect's face-threatening behaviour, the target is secretly recording the scene). If this were the case, the target would then be trying to protect her face by pretending that she is incapable of inferring the speaker's intended meaning. That would explain why she clings to the natural meaning of the suspect's utterances.

Interestingly enough, what we observe in Example 4 is an exciting verbal pulse between the suspect and the target. Because the target does not secure uptake of the suspect's intended meaning, the suspect fails to achieve his purpose (perlocutionary act) in the communicative situation. Whether the suspect's failure to secure uptake of the speaker's intended meaning is genuine or pretended, it is demonstrably true that, in this case, this passive resistance strategy is successful for the target.

Quid pro quo sexual harassment is more evident in exchange 49, in which the suspect is trying hard to negotiate the benefit he can get from the student in return for a pass on the exam. Significantly, the target asks the suspect to tell her what she has to do. In doing so, one may infer that the target is working from the assumption that it is the professor, the participant in the speech event "an exam" who has institutionally protected power, who must tell her what she has to do. (The target addresses the suspect with respect, using the *usted*[10] form of courtesy in Peninsular Spanish). In this way, she avoids performing a negative face-threatening act that would have caused damage to the suspect, but damages her own positive face by explicitly declaring her ignorance. The suspect cannot overtly say what he means because, if he did, his public social image could be damaged, and he could be held accountable for the crime of sexual harassment. Example 4 also illustrates the property of *reinforceability* that conversational implicatures have (Levinson 2000). Specifically, in exchange 49 the suspect overtly reformulates the *quid pro quo*, almost saying what he means as he implies that they must do something together: "*Pues si no hacemos nada, si no hacemos nada, otra cosa hija mía no puedo hacer.*" / "So if we don't do anything, if we don't do anything, another thing my darling I can't do."

7 Conclusions

This paper showed the relevance of a micro-language approach to the analysis of sexual harassment. Through the application of a pragmatics-based method, we demonstrated how pragmatics tools of linguistic analysis can help achieve an improved understanding of the covert nature of *quid pro quo* sexual harassment. For purposes of analysis, we turned to the linguistic toolbox, and selected several pragmatics tools. First, the pragmatic notion of speech event helped us to identify significant deviations from the mental frame, or schema, we have of "an exam." The results obtained in a small survey of 65 students confirmed the observed deviations. Then, we showed that these deviations were intentional, because they create a hostile environment by limiting the target's freedom of action and autonomy, thus causing damage to her face wants. Specifically, the macro-act of coercion emerges from repetitive behaviours that constitute emotional abuse for several reasons that Keashly (2001) has examined. Some of the reasons are: (a) the behaviours are unwelcome and unsolicited by the target, (b) the behaviours violate social norms and standards of conduct, as the perpetrator shows no respect for the target's dignity as a human being, (c) the behaviours result in harm or injury to the target, (d) the perpetrator intended to harm the target or could have refrained from the behaviour itself, (e) the perpetrator is

---

10 *Tú* and *usted* are both Spanish words for 'you.' *Tú* is informal and used when referring to someone who is a of the same status, a friend, or very young. By contrast, *usted* is formal and used when referring to someone who is of a higher status, an acquaintance, or simply older than the speaker.

in a more powerful position relative to the target. In the examples we analysed, the macro-act of coercion is linguistically realised through the repetition of face-threatening acts designed to humiliate and degrade the target, and to compel the target to act in a way contrary to her own interests. The threat of further harm – e.g. failing an exam – may lead to the cooperation or obedience by the person being coerced, although this did not happen in the case analysed here. Finally, it was possible to disclose the suspect's malicious intended meaning and conversational implicatures in the context of utterance. The suspect attempts to use his power over the target to compel her to engage in sexual activity. The vast majority of the students surveyed confirmed their assent that sexual coercion was present in this case. Precisely, under similar circumstances, 91% of the male students and 88% of the female students would think a suspect in such a situation was coercively soliciting sexual favours.

Determining the speaker's intent with any degree of certainty is not possible for any science, including linguistics, for the simple reason that researchers cannot peer into the minds of speakers or writers. Pointing out clues is as far as the expert linguist can go. It is up to the administrators to make judgements from there, and for the triers of fact to decide whether or not this was a case of *quid pro quo* sexual harassment. After two years, the administrators finally settled that the student had been a victim of *quid pro quo* sexual harassment. The length of time that it took to arrive at the administrative decision further illustrates the inherent difficulty in demonstrating cases of sexual harassment, even when the organisation is willing to combat this social scourge, and the victim heroically documents her account with a solid piece of physical evidence. In the example we analysed, the linguistic analysis of the audio recording – the only piece of evidence in the case – was essential for the administrators to understand the covert language of the perpetrator, and it smoothed the process of administrative decision-making that had been paralysed for two years.

References

Alba-Juez, Laura & Lachlan J. Mackenzie. 2016. *Pragmatics: cognition, context and Culture.* Madrid: McGraw Hill.
Austin, John L. 1962. *How to do things with words.* London: Clarendon.
Baron, Robert & Joel H. Neuman. 1998. Workplace aggression: The iceberg beneath the tip of workplace violence: Evidence on its forms, frequency, and targets. *Public Administration Quarterly* 21 (4). 446–464.
Bousfield, Derek. 2008. *Impoliteness and interaction.* Philadelphia & Amsterdam: John Benjamins.

Brown, Penelope & Stephen C. Levinson. 1987 [1978]. *Politeness: Some universals in language usage*. Cambridge, UK: Cambridge University Press.

Chamberlain, Lindsey Joyce, Martha Crowley, Daniel Tope & Randy Hodson. 2008. Sexual harassment in organisational context. *Work and Occupations* 35 (5). 262–295.

Council of Europe Convention on Preventing and Combating Violence against Women and Domestic Violence. 2011. https://www.coe.int/en/web/conventions/full-list/-/conventions/rms/090000168008482e (accessed 16 March 2021).

Cotterill, Janet (ed.). 2007. *The language of sexual crime*. London: Palgrave Macmillan.

Culpeper, John. 2011. *Impoliteness: Using language to cause offence*. Cambridge, UK: Cambridge University Press.

Dalmiya, Vrinda. 2002. Why is sexual harassment wrong? *Journal of Social Psychology*. 30 (1). 46–64.

Dick, Penny. 2013. The politics of experience: A discursive psychology approach to understanding different accounts of sexism in the workplace. *Human Relations* 66 (5). 645–669.

Galdi, Silvia, Anne Maass & Mara Cadinu. 2014. Objectifying media: their effect on gender role norms and sexual harassment of women. *Psychology of Women Quarterly* 38 (3). 398–413.

Grice, Herbert Paul. 1975. Logic and conversation. In Peter Cole and Jerry Morgan (eds.), *Syntax and Semantics 3: Speech acts*, 41–58. New York: Academic Press.

Goffman, Erving. 1967. On face work. In Erving Goffman (ed.), *Interaction ritual: Essays in face to face behavior*, 5–46. New York: Anchor Books.

Gumperz, John J. 1990. *Language and social identity*. New York: Cambridge University Press.

Hymes, Dell. 1974. *Foundations in sociolinguistics: An ethnographic approach*. Philadelphia: University of Philadelphia Press.

Keashly, Loraleigh. 2001. Interpersonal and systemic aspects of emotional abuse at work: The target's perspective. *Violence and Victims* 16 (3). 233–268.

Lakoff, Robin. 1973. The logic of politeness; or, minding your p's and q's. *Ninth regional meeting of the Chicago Linguistic Society*. Chicago: Chicago Linguistic Society.

Leech, Geoffrey. 1983. *Principles of pragmatics*. London: Longman.

Levinson, Stephen C. 2000. *Presumptive meanings: The theory of generalised conversational implicature*. Cambridge, MA & London: MIT Press.

Lutgen-Sandvik, Pamela & Sarah J. Tracy. 2012. Answering five key questions about workplace bullying: How communication scholarship provides thought leadership for transforming abuse at work. *Management Communication Quarterly* 26 (1). 3–47.

Malovich, Natalie J. & Jayne E. Stake. 1990. Sexual harassment on campus: Individual differences in attitudes and beliefs. *Psychology of Women Quarterly* 14. 63–81.

Nava, Mica. 2018. Sexual harassment, #MeToo and feminism. *Chartist for Democratic Socialism* 290. https://www.chartist.org.uk/sexual-harassment-metoo-and-feminism/ (accessed 11 March 2021).

Organic Law for Effective Equality between Women and Men (B.O.E. Nº 7123/03/2007). https://www.csd.gob.es/en/woman-and-sport/regulatory-framework/organic-law-effective-equality-between-women-and-men (accessed 28 November 2020).

Pandey, Kumar. 2009. Sexual harassment of working women. *A Journal of Asia for Democracy and Development* 9. 113–124.

Poyatos, Fernando. 1984. The multichannel reality of discourse. *Language Sciences* 6 (2). 306–337.

Scarduzio, Jennifer & Patricia Geist-Martin. 2010. Accounting for victimisation: Male professors' ideological positioning in stories of sexual harassment. *Management Communication Quarterly* 24 (3). 419–445.

Searle, John R. 1969. *Speech acts: An essay in the philosophy of language*. Cambridge, UK: Cambridge University Press.

Searle, John R. 1979. *Expression and meaning: Studies in the theory of speech acts*. Cambridge, UK: Cambridge University Press.

Searle, John R. 1983. *Intentionality: An essay in the philosophy of mind*. Cambridge, UK: Cambridge University Press.

Shor, Hilary M. 1992. Storytelling in Washington DC: Fables of love, power and consent in sexual harassment stories. *Southern California Law Review* 65 (3). 1347–1352.

Shuy, Roger W. 2012. *The language of sexual misconduct cases*. Oxford & New York: Oxford University Press.

Spencer-Oatey, Helen. 2005. (Im)politeness, face and perceptions of rapport: Unpackaging their bases and interrelationships. *Journal of Politeness Research: Language, Behaviour, Culture* 1 (1). 95–119.

Tannen, Deborah (ed.). 1993. *Framing in discourse*. Oxford & New York: Oxford University Press.

US Equal Opportunity Commission. Policy Guidance on Current Issues of Sexual Harassment. https://www.eeoc.gov/laws/guidance/policy-guidance-current-issues-sexual-harassment (accessed 12 October 2020).

US Equal Opportunity Commission. Title VII of the Civil Rights Act of 1964. https://www.eeoc.gov/statutes/title-vii-civil-rights-act-1964 (accessed 12 October 2020).

Watts, Richard J. 2003. *Politeness*. Cambridge, UK: Cambridge University Press.

Wright, Rita P. 2008. Sexual harassment and professional ethics. *SAA Archaeological Record* [Special issue] 8 (4). 27–30.

Stanisław Goźdź-Roszkowski
Hostility to religion or protection against discrimination? Evaluation and argument in a case of conflicting principles

1 Introduction

A number of recent legal cases have revolved around the responsibility for refusing to provide certain business services on the grounds of freedom of conscience, religion and speech.[1] Since the refusals involve clients from sexual minorities, these cases amount to a fundamental conflict of principles and values. The constitutionally protected freedoms of conscience, religion and speech clash with the equally protected right to non-discrimination and the protection of minority groups. Not surprisingly, such cases transcend national and jurisdictional borders, attracting considerable media attention and provoking very emotive responses in societies at large.

What merits particular attention is how judges account for their decisions in such hard cases. Supreme Court justifications seem to hold a special position, because they address a composite audience (Makau 1984) which, apart from participating litigants, Supreme Court justices, lower court justices, and the like, also includes different social groups interested in the outcome of specific civil rights cases. This chapter focuses on the construal of value-laden language and its role in judicial argumentation contained in the majority opinion of *Masterpiece Cakeshop, Ltd., et al. v. Colorado Civil Rights Commission et al.*[2] The Supreme Court of the United States in 2018 ruled in favour of a Colorado baker

---

[1] See, for example, *Brockie v. Dilinger* in Canada or *Bull & Bull v. Hall & Preddy* in Great Britain. In Poland, the Supreme Court in 2018 ruled against a print shop employee who refused to print banners for an LGBT business foundation, because he did not want to "promote" the gay rights movement (Case no. II KK 333/17 *Refusal to Provide Services on the Grounds of Freedom of Conscience and Religion*.)
[2] The text for analysis was accessed from https://www.oyez.org/cases/2017/16-111 (accessed 15 October 2020).

Acknowledgments: Research reported in this paper was supported by National Science Centre Poland under award number UMO-2018/31/B/HS2/03093.

Stanisław Goźdź-Roszkowski, University of Łódź

https://doi.org/10.1515/9783110720969-004

who had refused to create a wedding cake for a gay couple. The court's decision was narrow, and it left open the larger question of whether a business can discriminate against gay men and lesbians based on rights protected by the First Amendment. Instead, it addressed the question of whether the application of Colorado's public accommodations law, to compel a cake maker to design and make a cake that violates his sincerely held religious beliefs about same-sex marriage, violates the Free Speech or the Free Exercise Clauses of the First Amendment. The Supreme Court recognized the need to distinguish between conflicting legal principles and values: the rights and dignity of gay persons, and the right of all persons to exercise fundamental freedoms under the First Amendment (freedom of speech and the free exercise of religion).

In this contribution, I demonstrate how evaluative language and arguments are closely intertwined, and how evaluative language may contribute to the justificatory force of the arguments advanced by Supreme Court judges. For example, Justice Kennedy drafted the majority opinion referring to the Colorado Civil Rights Commission's treatment of the case as having "elements of a clear and impermissible hostility toward the sincere religious beliefs that motivated his [Masterpiece Cakeshop and Phillips] objection." The negative evaluation of the Colorado Civil Rights Commission's treatment of the case seems to lie at the heart of the court's decision to reverse the Commission's ruling against Masterpiece Cakeshop. The designation *evaluative language* is used to refer to the lexical level, where affective or evaluative word choice includes also the category of emotive words that are regarded as an extremely effective tool to persuade and encourage specific and desirable attitudes (Macagno & Walton 2014: 5). This means that the linguistic inquiry into evaluation is grounded within the argumentative and institutional realities of the legal case, by adding the perspective of legal argumentation theory (Dahlman & Feteris 2013; Feteris 2017).

To prepare the ground for the analysis and discussion, I will first shed more light on the concept of evaluative language and its relevance to legal justification (Section 2), and then I propose to adapt some concepts from the *pragmadialectical theory of argumentation* (Section 3) to view legal justification as a communicative and argumentative activity type. Section 4 presents the findings, while Section 5 brings the discussion and conclusions.

2 Evaluative language in legal justification

The linguistic perspective of studying evaluation bears some resemblance to the Attitudinal Model proposed in modern jurisprudence (Segal & Spaeth 2002). While both point towards the myth that judicial decisions are objective, impartial, and dispassionate, they also rely on the central concept of attitude, which comprises a relatively enduring and interrelated set of beliefs about an object or a situation. However, the linguistic approach will inevitably focus on how judges manifest their attitudes in writing. Huston and Thompson famously define evaluation as "the broad cover term for the expression of the speaker's or writer's attitude or stance towards, viewpoint on, or feelings about the entities or propositions that he or she is talking about" (2000: 5). Applied in the context of judicial writing, this means that it is only possible to analyse the actual verbal realization or manifestation of stances or attitudes included in the opinions of a given case.

While extremely complex in its application in real-life communicative contexts (Hunston 2010: 10–24), evaluation in discourse is usually studied in terms of a set of words and phrases which express evaluative meaning in a naturally occurring stretch of text. The focus is on identifying specific linguistic resources used to organise a discourse, or to organise the writer's stance towards either its content or the reader (Hyland 2000).[3] Evaluation can be understood as action performed in discourse, to refer to such situations when writers or speakers are actively engaged in expressing their stances.[4] The expression of attitude is then examined to determine how writers construe their subjective position and align themselves with their audiences. Different conceptualizations of evaluation open up numerous avenues for researching legal argumentation.[5] However, existing research focuses on isolating culturally and institutionally available generic resources, in an attempt to determine discursive strategies and language patterns adopted by judges to take stances as they organize their argumentative discourse (e.g. Goźdź-Roszkowski & Pontrandolfo 2013; Mazzi 2010).[6] For example, one distinctive pattern that has been identified in the literature consists of the so-called status-indicating nouns followed by a *that*-clause complement used in judicial opinions to signal sites of contention, i.e. challenged propositions are likely to be labelled as "arguments," "assumptions," "notions" or "suggestions" (Goźdź-Roszkowski 2018a). In the example below, taken from Justice Stevens' opinion of the Court, the proposition

---

[3] Evaluation can, of course, be made manifest at other levels of linguistic description: the phonological, morphological, syntactic, etc. See Alba-Juez and Thompson for a more detailed discussion (2014: 9–14).
[4] Note that some linguists use the term *stance-taking* for this understanding of evaluation (e.g. Englebretson 2007).
[5] See Goźdź-Roszkowski (2018b) for a more detailed overview of linguistic research into evaluation in judicial discourse.
[6] Depending on whether one looks at majority or dissenting opinions (e.g. Goźdź-Roszkowski 2020), the degree of subjectivity in the judicial voice might vary.

expressed as "that state courts must apply the restrictive Salerno test is incorrect as a matter of law" is labelled as "assumption," which may suggest that it is assessed as having a relatively weak epistemological value, because assumptions are not amenable to verification. Thus, evaluation can be seen working at the level of word-choice, since "assumption" was selected out of many different nouns potentially applicable in this context (e.g. assertion, claim, etc.) but also the specific qualification (the assumption is incorrect).

(1) Justice Scalia's assumption that state courts must apply the restrictive Salerno test is incorrect as a matter of law (*City of Chicago vs Morales et al.* 1999).

These studies tend to provide an inventory of lexical units or lexico-grammar patterns categorized according to their evaluative potential, and based on the pragmatic analysis of their immediate co-texts. The idea behind them is that the language used by judges is somehow constrained, and the writer operates within certain generic or institutional norms. While full of linguistic detail, such studies have relatively little explanatory power, i.e. they do not clarify why such evaluative items are used within the argumentative reality of legal justification, nor how they contribute to the realization of its argumentative goals.

This study shifts the analytical focus towards the link between the evaluative function of lexical items and the argumentative structure in the justification of judicial decisions. Perceived as the reasons and rationale given by courts in rendering their decisions, legal justification reflects the disciplinary and organizational culture of a given justice system. This means that the institutional framework and the corresponding legal form in which legal justifications are embedded may vary, depending on the legal system, jurisdiction, and type of court (e.g. appellate). Whereas judgments, as a whole, tend to have a fixed textual structure, with certain sections (e.g. headnote, procedural history, ruling or holding) usually being prepared by a court clerk, justifications instead reflect judicial reasoning. The language of justifications is inevitably less formulaic, and more likely to show idiosyncratic variation.

One fundamental characteristic that all justifications seem to have in common is the overarching goal of justifying the outcome of judicial decision-making process (Goźdź-Roszkowski 2020). Persuasive justification needs to be supported with sound argumentation (Feteris 2017), which in turn entails the analysis and evaluation of arguments advanced by various legal actors at different stages of court procedure. Such assessments are made against different standards of rational argument. The central question of what determines or contributes to the soundness of legal argumentation has been researched extensively by legal theorists, legal philosophers and argumentation theorists.[7] The concept of evaluation has remained at the forefront of these perspectives. For example, argumentation theorists have for a long time been interested in developing methods for the analysis and evaluation of legal argumentation (e.g. Walton 2016). They tend to assess the merits of legal argumentation based on certain norms of rationality, which serve as a basis for establishing whether or not an argument is sound and rational. However, argumentation scholars seldom pay attention to specific linguistic resources used by legal actors to construe their arguments. This contribution proposes to complement the existing linguistic and argumentative approaches to the analysis of legal justification by integrating the study of evaluative language into the argumentative reality of legal justification.

3 Pragma-dialectic approach to argumentation

In combining the study of evaluative language with argumentation, I propose to adapt two aspects of the *pragma-dialectical theory of argumentation* (van Eemeren 2018), which has been effectively applied to explore legal argumentation (Feteris 2015; 2012; Bletsas 2015). First, I use the ideal abstract model of a critical discussion to show how the instances of evaluative language and the different stages of the argumentation are closely intertwined to contribute to the realization of each stage. As Feteris aptly comments "legal justification is part of a dialogue, a *critical discussion* (my emphasis) aimed at the resolution of a dispute between the court and the multiple audience that must be convinced" (2017: 215). The critical discussion is carried at each stage by means of both dialectical and rhetorical components. While the former focuses on reasonableness, the latter aims for effectiveness. Second, I refer to the concept of strategic manoeuvring, originally proposed by van Eemeren (2010), as an attempt to reconcile the dialectical goal of resolving the difference of opinion with the rhetorical goal of steering the discussion in a desirable direction.

The ideal model of a critical discussion proposed in Feteris (2017: 206) specifies the stages which must be passed through to facilitate the resolution of a dispute, and the various speech acts which contribute to the process. In the *confrontation* stage the focal point of dispute is identified and the difference of opinion is established. In the *opening* stage, the participants reach agreement concerning the

---

[7] For an up-to-date survey of theories on the justification of judicial decisions, see Feteris (2017).

discussion rules, starting points and evaluation methods. In the *argumentation* stage, the point of view at issue is defended against critical reactions and the argumentation is evaluated; and in the *concluding* stage the final result is determined. It should be pointed out that the four-stage sequence does not always coincide with the chronological order of an actual discussion. For example, some stages could take place implicitly. While pragma-dialectics has already been used extensively with regard to judicial decisions, this chapter aims to extend the pragma-dialectic perspective with a linguistic one, by exploring how the use of evaluative language in the argumentative reality of actual cases is institutionalized with regard to the different stages of a critical discussion and the explicit and implicit conventions that govern the discussion.

4 The analysis

4.1 The confrontation stage

This stage of the critical discussion consists in realizing the dialectic goal of establishing the difference of opinion as well as the scope and content of the dispute. The introductory part of the opinion delivered by Justice Kennedy frames the dispute in terms of "difficulty." The case is assessed as "difficult" because it involves two legal principles which need to be reconciled: the protection of gay persons against discrimination and the right of all persons to exercise freedom of speech and religion:

(2) The case presents *difficult* questions as to the proper reconciliation of at least two principles. The first is the authority of a State and its governmental entities to protect the rights and dignity of gay persons who are, or wish to be, married but who face discrimination when they seek goods or services. The second is the right of all persons to exercise fundamental freedoms under the First Amendment, as applied to the States through the Fourteenth Amendment.

But there are other reasons for presenting the case as difficult, which are provided below in examples 3 to 5. One of the problems is linked to the concept of free speech and how the baker's cake fits this description.

(3) One of the *difficulties* in this case is that the parties disagree as to the extent of the baker's refusal to provide service.

(4) The free speech aspect of this case is *difficult*, for few persons who have seen a beautiful wedding cake might have thought of its creation as an exercise of protected speech.

(5) The same *difficulties* arise in determining whether a baker has a valid free exercise claim.

The strategic manoeuvring can be seen in the way the Court steers the discussion away from reconciling the conflicting principles and arriving at a more general and abstract principle. This is evidenced by the following two excerpts from the Court opinion:

(6) Given all these considerations, it is proper to hold that *whatever* the outcome of *some future* controversy involving facts similar to these, the Commission's actions *here* violated the Free Exercise Clause; and its order must be set aside.

(7) *Whatever* the confluence of speech and free exercise principles *might* be in *some* cases, the Colorado Civil Rights Commission's consideration of this case was inconsistent with the State's obligation of religious neutrality.

The use of "whatever" and the modal verb "might" signalling a remote possibility, as well as the spatial deictic *here*, all point toward the Court's reluctance to establish a precedent binding on sufficiently similar future cases. In effect, the court narrows the scope of its adjudication by focusing on how the case was treated by the Colorado Civil Rights Commission. There seems to be a link between the assessment of a case as "difficult" and the emphasis on the unique contexts in which legal controversies arise. The examples provided above are indicative of the pragmatic approach echoed in the following excerpt from *Bush v. Gore*: "Our consideration is limited to the present circumstances, for the problem [. . .] presents many complexities" (qtd. in Olsen 2017: 206, fn.2). This point will return in the concluding stage of Justice Kennedy's opinion, where numerous linguistic markers signal the restricted applicability of the court's ruling (emphasis in bold added):

(8) In **this** case the adjudication concerned a **context** that may well be **different** going forward in the respects noted above. However **later** cases raising **these or similar** concerns are resolved **in the future**, for these reasons the rulings of the Commission and of the state court that enforced the Commission's order must be invalidated.

(9) The outcome of cases like **this** in **other circumstances** must await **further** elaboration in the courts, all in the **context** of recognizing that **these** disputes must be resolved with tolerance, without undue disrespect to sincere religious beliefs, and without subjecting gay persons to indignities when they seek goods and services in an open market.

The Court is at pains to emphasize that the ruling given in this case is unlikely to be applicable beyond its originating context (cf. Olsen 2017: 206). In terms of surface linguistic features, this is manifested in the dense use of spatial and temporal deixis: the demonstratives ("this," "these"), forward-looking adjectives and adverbs ("further," "later," "in the future"). Worth noting is the use of modal verbs, especially the verb "may" in line 1 of example 8, where the justification seems to emphasize the possibility of reaching very different legal decisions in similar cases decided in the future. As a result, the scope of the ensuing discussion was clearly delimited.

4.2 The opening stage

According to the pragma-dialectic approach to legal argumentation, legal procedure is considered an institutional means to implement a critical discussion. Its aim is to resolve a dispute regarding the application of a specific legal rule in the context of a given case. This stage of the critical discussion consists in establishing common starting points and discussion rules (Feteris 2017). The Supreme Court chooses those starting points that are necessary to steer the critical discussion in the desired direction. On the one hand, it needs to give a positive evaluation of the petitioner's argumentation and a negative evaluation of the Colorado Civil Rights Commission's treatment of the case. On the other hand, the Court takes into account that acknowledging the claim of the petitioner is bound to raise issues related to discrimination against gay people. The following two excerpts from the Court's opinion provide an instance of pragmatic argumentation, which points to the consequences of applying a legal rule or principle (emphasis added):

(10) Yet if that exception were not confined, then a long list of persons who provide goods and services for marriages and weddings might refuse to do so for gay persons, thus **resulting in a community-wide stigma** inconsistent with the history and dynamics of civil rights laws that ensure equal access to goods, services, and public accommodations.

(11) And any decision in favour of the baker would have to be sufficiently constrained, lest all purveyors of goods and services who object to gay marriages for moral and religious reasons in effect be allowed to put up signs saying "no goods or services will be sold if they will be used for gay marriages," something that **would impose a serious stigma** on gay persons.

Both excerpts contain the explicitly negative and value-laden word "stigma" to argue what the likely outcome could be if the applicability of the legal principle is not limited. Excerpt (11) could in fact be treated as an instance of *argumentum ad absurdum*, because it refers to absurd and unacceptable consequences of refusing goods and services to gay couples if the scope of adjudication were not appropriately limited.

In another starting point, the Court takes into account different considerations which it weighs against each other, and it establishes the rule of religious neutrality as overarching in the consideration of this case (emphasis added):

(12) Our society has come to the recognition that gay persons and gay couples cannot be treated as social outcasts or as inferior in dignity and worth. For that reason the laws and the Constitution can, and in some instances must, **protect** them in the exercise of their civil rights. The exercise of their **freedom** on terms **equal** to others must be given great **weight** and **respect** by the courts. At the same time, the religious and philosophical objections to gay marriage are **protected** views and in some instances **protected** forms of expression. [. . .]

Nevertheless, while those religious and philosophical objections are **protected**, it is a **general rule** that such objections do not allow business owners and other actors in the economy and in society to deny **protected** persons equal access to goods and services under a neutral and generally applicable public accommodations law.

In (12), evaluation is expressed through a number of value-laden lexical items centring around the concepts of "freedom" and "protection." These could be interpreted as belonging to the category of emotive or ethical words (Macagno & Walton 2014), which introduce specific values into this early stage of the justification, and prepare the ground for the ensuing argumentation (cf. Perelman & Olbrechts-Tyteca 1973). The Court's starting position is that both parties' claims to freedom deserve protection. The assessments found in the passages indicate certain legal principles as fundamental starting-points. In Excerpt 12, it is the principle of equality which translates into the legal rule preventing businesses from refusing goods and services to protected people, i.e. people of a specific sexual orientation.[8] Yet, this rule, even if valid, was not regarded as a conclusive reason for the court's decision. Given the specific circumstances of this case, the principle of neutrality in considering the free speech claim gained the upper hand:

(13) But, nonetheless, Phillips was entitled to the neutral and respectful consideration of his claims in all the circumstances of the case.

4.3 The argumentation stage

In pragma-dialectic terms, the argumentation stage refers to the discussion strategy designed to reconcile the dialectic goal of determining the acceptability of argumentation advanced by a plaintiff (arguer) in light of the antagonist's attacks. In the context of this case, the Court focuses on a negative assessment of how the petitioner's case was handled by the Colorado Civil Rights Commission. Clearly, Justice Kennedy's opinion addresses arguments made by other legal actors (e.g. the Colorado Court of Appeal) but, arguably, the negative evaluation of the Colorado Civil Rights Commission's conduct remains at the core of the court's justification to reverse the judgment of the Colorado Court of Appeals. To illustrate how the "reasoning" component and the "effectiveness" component (outlined in Section 3) must be kept in balance in legal justification, I focus now on three "sites of evaluation" integrated into the court's argumentation, i.e. those portions of the justification that evaluate the Civil Rights Commission's treatment of the case, the petitioner's religious beliefs, and the nature of work performed by the petitioner.

As already mentioned, evaluating the Colorado Civil Rights Commission's treatment of the case is central to the Court's assertion that the petitioner's right to neutral and respectful consideration of his claim was violated. The Court's evaluation goes beyond a mere argument evaluation. On the one hand, it is broad, i.e. the evaluation target is not only an argument but also, more generally, the "public hearing" and "treatment" as in (15) and (16), respectively:

(14) The Civil Rights Commission's treatment of his case has some elements of a clear and impermissible hostility toward the sincere religious beliefs that motivated his objection.

---

[8] See Feteris (2017: 6) for a discussion of the distinction between legal rules and legal principles.

(15) That hostility surfaced at the Commission's formal, public hearings, as shown by the record.

(16) Another indication of hostility is the difference in treatment between Phillips' case and the cases of other bakers who objected to a requested cake on the basis of conscience and prevailed before the Commission.

The key word is "hostility," which is posited as the polar opposite of "sincere" as used in the phrase "sincere religious beliefs." Interestingly, Example 14 shows that evaluation is difficult to challenge if it is not the main point of the clause. This occurs when evaluation is presented in a clause as given. The phrase "sincere religious beliefs" is treated as given, not new information. In effect, the reader's acceptance of this evaluation is assumed rather than sought. The juxtaposition of "hostility" and "sincere" could be explained by the court's steering of the discussion towards recognizing the principle of "neutrality" as the desirable solution to the dispute in this case.

The evaluation is then made very specific by referring to the individual commissioner's statements. Justice Kennedy's opinion dwells on some instances of negative evaluation of the petitioner's religious beliefs, and it uses attributed evaluation, i.e. assigned to other voices (Partington 2013: 54–55), to add force to its argument that there is sufficient evidence of the Commission's hostility to religion. Examples of evaluative words are marked in bold:

(17) To describe a man's faith as "one of the most **despicable** pieces of rhetoric that people can use" is to **disparage his religion** in at least two distinct ways: by describing it as **despicable**, and also by characterizing it as **merely rhetorical** – something **insubstantial** and even **insincere**. The commissioner even went so far as to compare Phillips' invocation of his **sincerely held** religious beliefs to **defenses of slavery and the Holocaust**. This sentiment is **inappropriate** for a Commission charged with the solemn responsibility of **fair** and **neutral** enforcement of Colorado's antidiscrimination law – a law that protects discrimination on the basis of religion as well as sexual orientation.

Apart from the lexical items whose evaluative weight is intrinsic ("despicable," "insincere" or "insubstantial"), the negative evaluation is also expressed by reference to lexis that is primarily denotational, but which carries strong and negative associations ("slavery" and the "Holocaust"). Excerpt (17) is thus noteworthy for two main reasons.

First, it provides two layers of evaluation: evaluation attributed to the Commission and the averred evaluation by the Court. The former can be seen in how Justice Kennedy uses a direct citation to present the commissioner's point of view, and then how he reconstructs the commissioner's statement to demonstrate that the Commission has a clear bias against the bakery owner's Christian faith. Whether the words of the Commissioner should be indeed construed as expressing hostility to religion per se, rather than to the practice of invoking religion to serve one's own needs, is debatable.[9] The latter is found in a range of value-laden words ("disparaged his religion," "characterizing it as merely rhetorical") which represent what could be seen as "embedded evaluation," i.e. the Court's opinion evaluates the Commissioner's evaluation.

Second, certain nouns, referred to as "discourse nouns" (Francis 1994), can both contribute to the cohesion of a text and can simultaneously evaluate that text. In (17), the noun "sentiment" refers anaphorically to the preceding piece of text. It has the function of encapsulating the argumentation contained in the foregoing text, but it also evaluates it. The negative value judgment is signalled by the choice of the noun. The word "sentiment" refers to subjective belief and attitude, which may, in some contexts, seem excessive.[10] In addition, the noun is qualified negatively as "inappropriate."

One particular discursive strategy which projects evaluation as given information appears can be found in the court's opinion regarding faith. Below are two other instances corroborating the use of this discursive strategy, adopted in the justification to evaluate the petitioner's religious beliefs:

(18) The reason and motive for the baker's refusal were based on his sincere religious beliefs and convictions.

(19) As Phillips would see the case, this contention has a significant First Amendment speech component and implicates his deep and sincere religious beliefs.

---

9 See the critique of the court's argumentation in "Not a Masterpiece: The Supreme Court's Decision in *Masterpiece Cakeshop v. Colorado Civil Rights Commission*," by Erwin Chemerinsky (2017).

10 Evidence can be found in the numerous negatively-charged collocates of the word-form "sentiment," generated by the Collocate feature of the Corpus of Contemporary American English (COCA), at https://www.english-corpora.org/coca, such as *"anti-American," "strong," "popular," "anti-immigrant," "negative," "anti-Muslim."*

The reader is not positioned to decide whether or not she agrees with these evaluations. The reader's acceptance is assumed, and the subsequent argument is built upon that assumed acceptance. The proposition in (19) is, to some extent, dialogic, in that it opens up dialogic space for new ideas or for counterarguments. The use of the expression "As Philips would see the case" indicates that Justice Kennedy takes into account the possible existence of alternative viewpoints, in addition to the viewpoint he is advancing. In contrast, the statement in (18) is expressed in categorical and conclusive terms, fending off alternative views.[11]

As can be seen from the above presentation, the sites of evaluation can merge. The negative evaluation of how the Commission handled the case derives also from how the Commission appraised the petitioner's religious beliefs.

Finally, the third site of evaluation involves assessing the nature of work and service performed by the petitioner. The nature of the work performed by the petitioner is of crucial importance. As we recall in Example (4), one of the difficulties signalled in the justification is that of treating the petitioner's work in terms of an exercise of protected speech. To put it bluntly, can a cake be considered as an expression of speech that deserves protection? Indeed, this proved to be a contentious issue, leading to radically different assessments, as evidenced in Excerpt (20).

(20) Phillips raised two constitutional claims before the ALJ. He first asserted that applying CADA in a way that would require him to create a cake for a same-sex wedding would violate his First Amendment right to free speech by compelling him to exercise his artistic talents to express a message with which he disagreed. The ALJ rejected the contention that preparing a wedding cake is a form of protected speech and did not agree that creating Craig and Mullins' cake would force Phillips to adhere to "an ideological point of view." *Id.*, at

Interestingly, these conflicting assessments are attributed rather than averred by the Court. In the words of the Petitioner, his work is elevated to the status of artistic, unique and creative enterprise, which is tantamount to free speech and therefore deserves to be protected. In contrast, the assessment offered by the administrative law judge (ALJ) is devoid of any evaluative connotation, as it simply refers to "preparing a wedding cake." The court's justification does not

---

11 The difference between these two statements could be best explained by referring to the functional category of *engagement* in Martin and White's *The language of evaluation* (2005), where opening up the dialogical space is referred to as *expansion*, while fending off alternative views is called *contraction*.

overtly endorse either evaluation. While recognizing the difficulty of treating the creation of a beautiful wedding cake as an exercise of protected speech, the court observes that "this is an instructive example, however, of the proposition that the application of constitutional freedoms in new contexts can deepen our understanding of their meaning." Yet, the court's positive evaluation could be construed as being made implicitly by not questioning the petitioner's assertions:

(21) Phillips claims, however, that a narrower issue is presented. He argues that he had to use his artistic skills to make an expressive statement, a wedding endorsement in his own voice and of his own creation.

In other words, the cake is the result of the Petitioner's creative skills, capable of sending a meaningful message. The effect of the court's assessment is significant, as it extends the notion of free speech.

4.4 The conclusion stage

This stage brings the result of the critical discussion. The judge determines which of the different positions in the dispute is justified, based on the common legal and factual starting points (Feteris 2017: 206). The judge's decision should be seen in light of how the Supreme Court manoeuvres strategically to avoid addressing the underlying issue: whether or not a business has a constitutional right to discriminate based on its owner's religious beliefs. Instead, the justification focuses on narrower grounds, by concluding that members of the Colorado Civil Rights Commission expressed impermissible hostility to religion:

(22) While the issues here are difficult to resolve, it must be concluded that the State's interest could have been weighed against Phillips' sincere religious objections in a way consistent with the requisite religious neutrality that must be strictly observed. The official expressions of hostility to religion in some of the commissioners' comments – comments that were not disavowed at the Commission or by the State at any point in the proceedings that led to affirmance of the order – were inconsistent with what the Free Exercise Clause requires.

It is the word "consistency" that is the prime lexical carrier of evaluation, signalling the use of arguments of coherence to show that the decision reached in this case does not fit with the legal rule enshrined in the Free Exercise Clause.

5 Discussion and conclusions

The pragma-dialectical theory of argumentation and discourse analysis presented here enabled us to observe a broad intersection of evaluative language with the argumentative discourse found in the Supreme Court's justification. In fact, it is argued that evaluative language permeates argumentative discourse. This is not surprising, since evaluation is inherent in legal justification. Judges use evaluative language to assess arguments advanced by other legal actors, but they resort to it to support their own lines of argumentation.

In this contribution, the ideal model of a critical discussion was used as a tool to describe the ways in which argumentation is substantiated in the institutional macro-context of Supreme Court justification. The focal points – the use of evaluative language, also referred to as "sites of evaluation" – have been characterized as corresponding to the stages of a critical discussion. These include the initial situation, the starting points, the argumentative means, and the outcome of the exchange. This approach has helped expand the methodological perspective of studying evaluative language in discourse communication, by adding the context of a real-life communicative activity connected with the institutionalized context of legal justification. It turns out that, in line with previous research on the justification of judicial decisions, the communicative activity type (legal procedure) imposes certain constraints on the strategic manoeuvring and the argumentative patterns that are found in this particular institutional context to resolve the difference of opinion.

The concept of strategic manoeuvring was adopted to account for the way the Supreme Court combined the rational resolution of the legal dispute, which boiled down to the fundamental tension between equality and liberty, and to the rhetorical choice and presentation of discussion moves. The rhetorical choice and the "presentational devices" (Eemeren 2010) include value-laden language, selected with a view to steering the discussion towards the conclusion which left unresolved the key question of whether forcing businesses to provide services for gays and lesbians, or others, violates free exercise of religion or free speech rights of owners who wish to refuse to provide such services.

In the confrontation stage of the justification, the Supreme Court frames the dispute as "difficult" – in Dworkinian terms, a hard case (Dworkin 1986). This assessment paves the way for limiting the applicability of the final ruling, and for shying away from giving an unequivocal answer to the question of whether a business's freedom to choose its customers is more important than preventing discrimination on the grounds of sexual orientation. However, even if the Court did not resolve this issue, it did indicate that claims similar to that of Jack Phillips and Masterpiece Cakeshop were unlikely to prevail under the free exercise clause. Justice Kennedy's opinion suggested that the free exercise clause would not provide a basis for such refusals of service when there is no specific expression of hostility to religion. The Court clearly stated: "while those religious and philosophical objections are protected, it is a general rule that such objections do not allow business owners and other actors in the economy and in society to deny protected persons equal access to goods and services under a neutral and generally applicable public accommodations law." This is important if we consider that, in the opening stage, the Supreme Court establishes the common legal starting points in such a way that it can focus on the negative evaluation of the Colorado Civil Rights Commission arguments.

The Court's evaluation of the Commission's conduct as hostile underpins much of the argumentative stage. The Court proceeds to provide evidence as demonstrating impermissible hostility to religion by the Colorado Civil Rights Commission. It is interesting to note that this point in the Court's justification is based merely on the opinion of one of the commissioners that people had often committed acts of blatant discrimination in the name of a religion, and that no religion should be used to hurt others. A statement that, sadly, could be easily proved as factually correct. As Justice Ginsburg, joined by Justice Sotomayor in dissent, explained, "Whatever one may think of the statements in historical context, I see no reason why the comments of one or two Commissioners should be taken to overcome [the baker's] refusal to sell a wedding cake" to the same-sex couple. But then, the Court chose to construe the Commissioner's words as an expression of hostility to religion, which is crucial in light of the rhetorical goal of steering the discussion in the desirable direction.

Equally central are the descriptions of the nature of the petitioner's work and his religious beliefs. To appraise baking a cake as "expressive activity" results in accepting that forcing the petitioner to prepare a cake for the gay wedding constitutes impermissible compelled speech, in violation of the First Amendment. The repeated references to Jack Philips "sincerely held" religious belief in the Court's justification could reflect that the tendency is for the US Supreme Court to focus in religious freedom cases on whether a particular person does actually have a sincerely held religious belief, rather than on what the religion teaches. In Poland, the Supreme Court disputed the claim of a print shop employee who refused to print banners for an LGBT business foundation, citing his Roman Catholic faith.[12] The court pointed to the Catechism of the Catholic Church (section 2358) which

---

[12] The case was decided by Poland's Supreme Court in 2018 (Case no. II KK 333/17 *Refusal to Provide Services on the Grounds of Freedom of Conscience and Religion* (see Goźdź-Roszkowski, forthcoming).

says that "They (gay people) must be accepted with respect, compassion, and sensitivity. Every sign of unjust discrimination in their regard should be avoided." The Court concluded that citing religious beliefs, which were in fact based on his subjective feelings concerning the LGBT people, did not constitute reasonable and justified reason for refusing to provide the service.

The approach adopted in this study has attempted to bring together two aspects of legal justification: rational dispute resolution and a rhetorical orientation. It is the latter that relies most on evaluative language to steer the discussion in a particular direction perceived as desirable by the court. While evaluative language never constitutes an argument per se, in a pragma-dialectic reconstruction of a legal decision, it helps to shed light on the complexities of a legal case by focusing our attention on argument construction and use.

References

Alba-Juez, Laura & Geoff Thompson. 2015. The many faces of evaluation. In Geoff Thompson & Laura Alba-Juez (eds.), *Evaluation in context*, 3–23. Amsterdam: John Benjamins.

Bletsas, Marina. 2015. The voices of justice: Argumentative polyphony and strategic manoeuvring in judgement motivations. In Frans H. van Eemeren & Bart Garssen (eds.), *Scrutinizing argumentation in practice* 9, 80–97. Amsterdam: John Benjamins.

Chemerinsky, Erwin. 2017. Not a masterpiece: The supreme court's decision in Masterpiece Cakeshop v. Colorado civil rights commission. *Human Rights* 43 (4). Online. https://www.americanbar.org/groups/crsj/publications/human_rights_magazine_home/the-ongoing-challenge-to-define-free-speech/not-a-masterpiece/ (accessed 25 March 2021).

Dahlman, Christian & Eveline Feteris. 2013. *Legal argumentation theory: Cross-disciplinary perspectives*. Dordrecht, Boston & Lancaster: Springer.

Dworkin, Ronald. 1986. *Law's empire*. London: Fontana.

Eemeren, Frans H. van. 2010. *Strategic manoeuvring in argumentative discourse: Extending the pragma-dialectic theory of argumentation*. Amsterdam: John Benjamins.

Eemeren, Frans H. van. 2018. Strategic manoeuvring in argumentative discourse. In *Argumentation theory: A pragma-dialectical perspective*. Dordrecht, Boston & Lancaster: Springer.

Englebretson, Robert (ed.). 2007. *Stancetaking in discourse: Subjectivity, evaluation interaction*. Amsterdam: John Benjamins.

Feteris, Eveline T. 2015. The role of pragmatic argumentation referring to consequences, goals, and values in the justification of judicial decisions. In Bart Garssen, David Godden, Gordon Mitchell & Arnolda Francisca Snoeck Henkemans (eds.), *International society for the study of argumentation (ISSA): 8th international conference on argumentation: July 1–July 4, 2014*, 414–425. Amsterdam: Sic Sat.

Feteris, Eveline T. 2017. *Fundamentals of legal argumentation: A survey of theories on the justification of judicial decisions*. Dordrecht, Boston & Lancaster: Springer.

Feteris, Eveline T. 2012. Strategic manoeuvring in the case of the "unworthy spouse." In Frans H. van Eemeren & Bart Garssen (eds.), *Exploring argumentative contexts*, 149–164. Amsterdam: John Benjamins.

Francis, Gill. 1994. Labelling discourse: An aspect of nominal-group lexical cohesion. In Malcolm Coulthard (ed.), *Advances in written text analysis*, 83–101. London: Routledge.

Goźdź-Roszkowski & Gianluca Pontrandolfo. 2013. Evaluative patterns in judicial discourse: a corpus-based phraseological perspective on American and Italian criminal judgments. *International Journal of Law, Language and Discourse*, 3 (2). 9–69.

Goźdź-Roszkowski, Stanisław. 2018a. Between corpus-based and corpus-driven approaches to textual recurrence: Exploring semantic sequences in judicial discourse. In Kopaczyk Joanna & Jukka Tyrkko (eds.), *Applications of pattern-driven methods in corpus linguistics*, 131–158. Amsterdam: John Benjamins.

Goźdź-Roszkowski, Stanisław. 2018b. Counting the uncountable? Quantitative and qualitative methods of analysing evaluative language in institutional discourse: A corpus linguistics perspective. In Łukasz Bogucki & Piotr Cap (eds.), *Explorations in language and linguistics*, 145–158. Bern: Peter Lang.

Goźdź-Roszkowski, Stanisław. 2020. Move analysis of legal justifications in constitutional tribunal judgments in Poland: What they share and what they do not. *International Journal for the Semiotics of Law* 33. 581–600.

Goźdź-Roszkowski, Stanisław. Forthcoming in 2022. Evaluative language and strategic manoeuvring in the justification of judicial decisions: The case of teleological-evaluative argumentation. In Stanisław Goźdź-Roszkowski & Gianluca Pontrandolfo (eds.), *Law, language and the courtroom: Legal linguistics and the discourse of judges*. London: Routledge.

Hunston, Susan. 2010. *Corpus approaches to evaluation, phraseology and evaluative language*. London: Routledge.

Hyland, Ken. 2000. *Disciplinary discourses: Social interactions in academic writing*. Detroit: University of Michigan Press.

Macagno, Fabrizio & Douglas Walton. 2014. *Emotive language in argumentation*. Cambridge, UK: Cambridge University Press.

Makau, Josina M. 1984. The Supreme Court and reasonableness. *Quarterly Journal of Speech* 70 (4). 379–396.

Martin, James R. & Peter R. White. 2005. *The Language of evaluation: Appraisal in English*. Basingstone: Palgrave Macmillan.

Mazzi, Davide. 2010. 'This Argument Fails for Two Reasons . . . ': A linguistic analysis of judicial evaluation strategies in US Supreme Court Judgements. *International journal for the Semiotics of Law* 23 (4). 373–385.

Olsen, Frances. 2017. Pragmatic interpretation by judges: Constrained performatives and the deployment of gender bias. In Janet Giltrow & Dieter Stein (eds.), *The pragmatic turn in law*, 205–232. Boston & Berlin: De Gruyter Mouton.

Partington, Alan, Duguid Alison & Charlotte Taylor. 2013. *Patterns and Meanings in Discourse. Theory and Practice in Corpus-Assisted Discourse Studies (CADS)*. Amsterdam: John Benjamins.

Perelman, Chaim & Lucie Olbrechts-Tyteca. 1973. *The new rhetoric: A treatise on argumentation*. Notre Dame, IN: University of Notre Dame Press.

Pontrandolfo, Gianluca & Stanisław Goźdź-Roszkowski. 2014. Exploring the local grammar of evaluation: The case of adjectival patterns in American and Italian judicial discourse. *Research in Language* 12 (1). 71–92.

Segal, Jeffrey & Harold Spaeth. 2002. *The Supreme Court and the attitudinal model revisited.* Cambridge, UK: Cambridge University Press.

Thompson, Geoff & Susan Hunston. 2000. Evaluation: An Introduction. In Susan Hunston & Geoff Thompson (eds.), 1–27. *Evaluation in text: Authorial stance and the construction of discourse.* Oxford: Oxford University Press.

Walton, Douglas. 2016. *Argument evaluation and evidence.* Dordrecht, Boston & Lancaster: Springer.

James Vanden Bosch
Heller (2008) and the language of the Second Amendment: Grammar, meaning, and canonical conventions

1 Preamble

Legal scholars use linguistic tools in their work as a matter of course, and scholars of Constitutional law regularly engage in semantic analysis, trying to determine what specific words and phrases mean in context. Historical linguistics (the study of how language changes over time) and corpus linguistics (the analysis of large bodies of text, now typically aided by computers) also have roles to play in the study of Constitutional law, and these tools prove to be particularly helpful in the study of the 2008 Supreme Court decision on the Second Amendment, *District Of Columbia, et al., Petitioners v. Dick Anthony Heller* (2008) (hereafter referred to as *Heller*). One such subject of discussion in *Heller* is the meaning of "to bear arms" in the main clause; I'll consider that issue later in this paper. The other matter has to do with the opening phrase of the Second Amendment, and its grammatical and semantic relationship to the main clause.

The opening phrase of the Second Amendment of the U.S. Constitution is almost certainly the most famous and most consequential absolute phrase in American legal history. This is the text of the Second Amendment, as ratified in 1793: "A well-regulated militia being necessary to the security of a free state, the right of the people to keep and bear arms shall not be infringed." In the legal analysis of this text, the amendment is ordinarily divided into two parts:

> (1) the **prefatory clause or material** – "A well-regulated militia being necessary to the security of a free state" and

> (2) the **operative clause** – "the right of the people to keep and bear arms shall not be infringed."

The corresponding grammatical analysis uses a different set of labels:

> (1) an **absolute phrase** – "A well-regulated militia being necessary to the security of a free state" and

---

James Vanden Bosch, Calvin University

https://doi.org/10.1515/9783110720969-005

(2) an **independent (or main) clause** – "the right of the people to keep and bear arms shall not be infringed."

There are many late-18th-century grammatical variations on absolute phrases that feature the verb "being"; here are a few of the most common variations that do not have the same grammatical structure as the prefatory material of the Second Amendment (the absolute phrase in **bold**):

(1) S–V with no subjective complement

1795: George Washington: Official letters to the Honourable American Congress:

The <u>intelligence</u> transmitted by general Arnold <u>being</u> of an extremely interesting and important nature, I thought it advisable to forward the same immediately by express.

(2) Expletive (there)–V–delayed subject

1794: American state papers: public lands vol.1:

There <u>being</u>, however, a considerable <u>number</u> of unlocated warrants still extant, held as well by minors as others, who are either ignorant of the limitation, or whose interest has been neglected by those who represented them, and the object of the limitation being not to preclude, but to hasten the location of those warrants, both justice and policy require that there should be a further extension of the time.

(3) S–LV–predicate noun

1794: American state papers: public lands vol. 1:

There being, however, a considerable number of unlocated warrants still extant, held as well by minors as others, who are either ignorant of the limitation, or whose interest has been neglected by those who represented them, and **the <u>object</u> of the limitation <u>being</u> not <u>to preclude, but to hasten the location of those warrants</u>**, both justice and policy require that there should be a further extension of the time.

(4) Expletive (it)–V–delayed subject

1781: Order in council concerning the copying of public records, [10 May 1781]; Virginia executive council:

The letters and other Papers of the Council having been destroyed in the expedition of the enemy to the Town of Richmond in the month of January last, and **it being of general importance that memorials of public events be preserved**, and particularly interesting to those having a share in the administration that the records of their proceedings should under every possible circumstance guard them against misrepresentation and mistake and the board being of opinion that copies may be obtained of many letters and other papers of considerable importance by application to those to and from whom they have been written, Advise that a proper person be appointed to execute this business, that he be instructed particularly to go to Congress and to General Washington in order to obtain permission to copy the letters which have passed between them and this board previous to the commencement of the present year –

(5) S–passive-voice verb

1774: Journals of the continental congress:

A motion was made by [Mr. William] Pierce, seconded by Mr [William] Few repealing the order passed Sept P' 1786, and on the Question to agree to the motion, **the Yeas and Nays being required** by Mr [William] Pierce, 1 May 9, 1787.

All five structures in bold, above, are absolute phrases, but they do not have the same grammatical structure as the prefatory material of the Second Amendment.

The grammatical structure of that absolute phrase is as follows:

Subject	Incomplete form of linking verb	Predicate adjective
A well-regulated <u>militia</u>	**being**	**<u>necessary</u> to the security of a free state**

There is also an *It*-cleft[1] version of such absolute phrases, in which the "It" takes the subject position and is followed by "being," which is in turn followed by the predicate adjective and then by the grammatical subject of the phrase:

(6) It-cleft–LV–PA–delayed subject

1781: From Thomas Jefferson to the county lieutenant of Henrico, 19 April 1781:

---

[1] In it-cleft structures, or it-extrapolation, part of the sentence is moved from its expected place, and placed at the front. For example, "I leave you with regret" becomes "It is with regret that I leave you." Here "regret" moves from end of sentence to front of the sentence, as complement for <is>.

> Sir, **It being possible that the enemy may be destined for this place** we have advised the Auditors, Treasurer, Register, Clerks of the Assembly Chancery and General Court, the Clothier and Commissary of Naval Stores to prepare their Papers & c. for Removal.

2 Originalism and *Heller* (2008)

Even with the different naming protocols, the text of the Second Amendment is the same sentence for grammarians, for constitutional lawyers, and for Supreme Court justices. Justice Antonin Scalia insisted that this was the case; in the general interpretive principle announced in *District of Columbia v. Heller* (2008), Justice Scalia stated his working assumption as an originalist this way:

> In interpreting this text, we are guided **by the principle that "[t]he Constitution was written to be understood by the voters; its words and phrases were used in their normal and ordinary as distinguished from technical meaning**." United States v. Sprague, 282 U. S. 716, 731 (1931); see also Gibbons v. Ogden, 9 Wheat. 1, 188 (1824). Normal meaning may of course include an idiomatic meaning, but it excludes secret or technical meanings that would not have been known to ordinary citizens in the founding generation".
> (*Heller* 2008: 3; my emphasis)

Justice Scalia's statement of this principle is an open invitation to others to work with the same assumptions, and to ask the same basic question when considering the twenty-seven words of the Second Amendment: What would American voters, ordinary citizens in the founding generation, have understood these words and phrases to mean, as used in their normal and ordinary meanings?

3 The Corpus of Early Modern English (CEME) and the Corpus of Founding Era American English (COFEA), 2018

The *Heller* decision was a startling development in the history of the interpretation of the Second Amendment, because the standard interpretation had been one in which the absolute phrase at the front established the narrow context in which the main clause existed – it protects the right to bear arms for those who serve in the state-run well-regulated militias of the day. The *Heller* decision (with a 5–4 majority) claimed that the prefatory material was not allowed to broaden or narrow the meaning of the language of the main clause, and it also argued that, in the late 18th century, the phrase "to bear arms" referred to much more than a military use of firearms – it also meant the use of firearms for sport and self-defence.

The arguments about the meanings of the words that make up the absolute phrase and the main clause of the Second Amendment haven't diminished since the 2008 *Heller* opinion, but there has been a significant development in the analytical tools available to grammarians and legal scholars; courtesy of the Brigham Young University Law School, two new corpora became available early in 2018: the Corpus of Early Modern English (1475–1800; 1,107,365,393 words); and the Corpus of Founding Era American English (1760–1799; 136,848,583 words).

Many readers of this volume are already aware of an early use of these corpus resources in relationship to the *Heller* opinion. On May 21, 2018, Dennis Baron published a short essay in *The Washington Post* with the title "Antonin Scalia was wrong about the meaning of 'bear arms,'" and the argument he presented there is based on his research using these two corpora. Dennis Baron used these two corpora to provide more context for the meaning of the phrase "bear arms" in the main clause of the Second Amendment. Although the *Heller* opinion argued for several meanings of this phrase, including non-military meanings, for hunting and for self-defence, Baron's analysis of approximately 1500 uses of that phrase provided quite a different emphasis. This is Justice Scalia's conclusion in *Heller* regarding the meaning of "bear arms" in the Second Amendment:

> Although ['bear arms'] implies that the carrying of the weapon is for the purpose of 'offensive or defensive action,' it in no way connotes participation in a structured military organization. From our review of founding-era sources, we conclude that this natural meaning was also the meaning that 'bear arms' had in the 18th century. In numerous instances, 'bear arms' was unambiguously used to refer to the carrying of weapons outside of an organized militia.

And this is Dennis Baron's response:

> A search of Brigham Young University's new online Corpus of Founding Era American English, with more than 95,000 texts and 138 million words, yields 281 instances of the phrase 'bear arms.' BYU's Corpus of Early Modern English, with 40,000 texts and close to 1.3 billion words, shows 1,572 instances of the phrase. Subtracting about 350 duplicate matches, that leaves about 1,500 separate occurrences of 'bear arms' in the 17th and 18th centuries, and only a handful don't refer to war, soldiering or organized, armed action. These databases confirm that the natural meaning of 'bear arms' in the framers' day was military.

4 The absolute phrase in *Heller*

Dennis Baron's engagement with Justice Scalia's reading of "bear arms" is not the focus of my paper, but Baron's work with these two corpora makes it clear that the *Heller* opinion is not the last word on the interpretation of the words and phrases of the Second Amendment "in their normal and ordinary [. . .] meaning." My focus in this paper is on Justice Scalia's argument regarding the semantic relationship between the absolute phrase and the main clause. Early on in *Heller*, Justice Scalia describes the relevance of the absolute phrase (which he refers to below as the "prefatory clause"):

> The Second Amendment is naturally divided into two parts: its prefatory clause and its operative clause. **The former does not limit the latter grammatically, but rather announces a purpose.** (*Heller* 2008: 3; my emphasis)

The first part of the highlighted sentence is a reference to an interpretive canon that I'll describe more fully in section X. But the second part of this sentence concedes that the prefatory material, the absolute phrase, "announces a purpose," namely the preservation of the militias so that the militias can maintain and protect national security. Much later in *Heller*, as a kind of summary description, Scalia addresses the meaning and relevance of the absolute phrase in context again:

> It is therefore entirely sensible that the Second Amendment's prefatory clause announces the purpose for which the right was codified: to prevent elimination of the militia. **The prefatory clause does not suggest that preserving the militia was the only reason Americans valued the ancient right; most undoubtedly thought it even more important for self-defence and hunting.** (*Heller* 2008: 26; my emphasis)

In both passages from *Heller*, the more or less common-sense interpretation of the prefatory material is that the absolute phrase announces the purpose of the Second Amendment: the protection of the militias in order to maintain the security of the state. But in both quoted passages from *Heller*, that common-sense reading is presented only to be undercut; Justice Scalia asserts that the absolute phrase does not govern the operative clause and he asserts that the framers of this amendment must have had other, even more important, reasons to secure this right to keep and bear arms, namely, self-defence and hunting. He argues that the explicit purpose stated in the absolute phrase cannot govern the main clause, and he also asserts that two reasons not mentioned in the text of the amendment must have been more important for the drafters of the Second Amendment than the preservation of the militia was. It is this unusual reading of the semantics of the Second Amendment that brought me to CEME and COFEA, so that I could learn more about how such phrases and clauses were used during the Federalist era.

5 "Being necessary" absolute phrases in CEME and COFEA

In my initial work with these two corpora, I searched for absolute phrases in the Federalist era that are similar in structure and sense to the absolute phrase used in the Second Amendment. I searched both corpora, not for absolute phrases in general, but for absolute phrases constructed with an expressed subject that had "being" as its incomplete linking verb and "necessary" as the predicate adjective of the phrase.

For the British context, I searched the Corpus of Early Modern English (CEME); here are the results, in several stages:

CEME, for the entire corpus, 1475–1800: 421 "being necessary" absolutes
CEME, only the Evans file of American sources: 86 "being necessary" absolutes
CEME, British sources only: 335 "being necessary" absolutes
CEME, British sources only, 1760–1799: 109 "being necessary" absolutes

I isolated the British absolutes count for the years 1760 to 1799 because COFEA deals with the same time span. I then searched the Corpus of Founding Era English (COFEA), made up of writing published in America from 1760 to 1799. This particular absolute phrase was approximately 2.66 times more common in America during this period than it was in Great Britain. Here are the results of my COFEA search for all instances of absolute phrases made up of a subject, "being" as the verb form, and "necessary" as the predicate adjective:

1760s: 28 "being necessary" absolutes
1770s: 83
1780s: 64
1790s: 115
Total, 1760–1799:   290 "being necessary" absolutes

In what follows, I'll focus solely on the absolute phrases extracted from COFEA.

6 Grammatical features of the 290 absolute phrases from COFEA

The "being necessary" absolute phrases from COFEA come in two grammatical varieties:

(1) The standard variety: Subject–being–necessary:

                S   LV   PA

A well-regulated **militia being necessary** to the security of a free state, the right of the people to keep and bear arms shall not be infringed.

(2) The *It*-cleft variety: Expletive (It)–being–necessary–delayed subject:

Expl LV   PA   Infinitive phrase functioning as the delayed subject
It being necessary to provide a stronger bridle for unquiet and seditious spirits, a new and most rigorous law must be made to beat down their arrogance and insolence.

Note: An adverb can appear between the V and the PA: S–V–ADV–PA:

        S   V   ADV   PA
A large **supply** of sheets, blankets, & c. **being immediately necessary**, Thomas Wistar and Henry Deforest are requested to take their instructions from the managers and procure every necessary that may be required for the use of the sick at the Hospital.

These grammatical varieties turn up in the results of my COFEA research as follows:

S–LV–PA variety: 153 instances, including 40 with the adverb insertion
It-cleft variety: 137 instances, including 32 with the adverb insertion
Total: 290 instances of the "being necessary" absolute phrase

7 The placement of absolute phrases in the COFEA extractions

The 290 "being necessary" absolute phrases appear in three positions in relationship to the main clauses that they modify: initial, medial, and final. Here are several examples from COFEA:

Initial Position: "**A well-regulated militia being necessary to the security of a free state**, the right of the people to keep and bear arms shall not be infringed."

Medial Position: "Nothing will recover them but the vigorous exertions of Men of abilities who know our wants, & the best means of supplying them – these Sir without a compliment I think you possess – why then, **the department being necessary**, should you shrink from the duty of it."

Final position: "They require water, **it being necessary to overflow the grounds where they are cultivated**."

Of the 290 absolute phrases I extracted, 183 are in the initial position, 8 are in the medial position, and 99 are in the terminal position.

Summary: COFEA extractions of "being necessary" absolute phrases by grammar and placement:

Decade	Instances	Grammar		Placement of absolute phrase		
		S–V–PA	*It*-cleft	Initial	Medial	Final
1760s	28	3	25	24	0	4
1770s	83	33	50	51	1	31
1780s	64	39	25	37	4	23
1790s	115	78	37	71	3	41
Total	290	153	137	183	8	99

8 Extractions under consideration

Because the standard-variety "being necessary" absolutes (S–V–PA) are identical in grammatical form and word-order to the absolute phrase that introduces the Second Amendment, and because the initial-placement absolutes follow the placement of the absolute phrase in the Second Amendment, I will initially focus my analysis on the 55 sentences from COFEA that have the identical pattern: all of them are S–V–PA absolute phrases placed before the clauses that they modify.

I decided to focus on these 55 sentences not because the others are not worth examining, but because (1) these 55 are the best matches for the grammatical structure used in the Second Amendment; and (2) the Prefatory-Materials Canon applied by Justice Scalia in *Heller* requires or assumes that the prefatory material is in the initial position. Of the 153 standard-variety absolute phrases I extracted, 69 are in the initial position. In that group of 69 are 14 duplicates, including 12 instances of the text of the Second Amendment itself. (See Appendix I, below, for the 55 sentences under consideration).

9 The uses and meanings of these "being necessary" absolute phrases in context

Some of the 55 passages I extracted from COFEA follow similar but familiar patterns: a statement of a principle followed by an application of that principle; a statement of a problem or need followed by a solution; or a law and its consequences. Because of the order of these statements – a "being necessary" absolute phrase followed by the operative clause – there is a recognizable pattern that makes a direct connection between the prefatory material and the operative clause.

A closer look at several of these passages should make this relationship obvious:

(1) 1799: From Alexander Hamilton to James Miller, 22 October 1799; letter: official
correspondence:

> **The axes, Cross cut saws, files and Grindstones, being necessary to begin the work** they must not be delayed for the purpose of procuring the others.

Here, the project at hand will require cutting and shaping tools and the means to sharpen them, and because the cutting and shaping work must come first, it is important for the person in charge of ordering and delivering these tools to procure them first – no other tools are needed to begin the project, and no other tools needed later in the project can be put to use until this initial work has been completed. Order the shaping tools now, and make sure that they arrive first.

(2) 1794: Minutes of the proceedings of the committee, appointed on the 14th September, 1793, by the citizens of Philadelphia, the Northern Liberties and the District of Southwark, to attend to and alleviate the sufferings of the afflicted with the malignant fever, prevalent, in the city and its vicinity, with an appendix.; committee minute:

A large supply of sheets, blankets, & c. being immediately necessary, Thomas Wistar and Henry Deforest are requested to take their instructions from the managers and procure every necessary that may be required for the use of the sick at the Hospital.

The purpose announced here is the immediate acquisition of sheets and blankets and other basic hospital supplies for dealing with the outbreak of malignant fevers in Philadelphia; the main clause then specifies how this purpose will be accomplished, namely, by the work of Wistar and Deforest in cooperation with the managers.

(3) 1799: An oration pronounced at Claremont, on the anniversary of American independence. July 4th, 1799. By Samuel Fiske, A.M.; patriotic oration:

SOCIETY being necessary for the welfare and happiness of man; so government and subordination are necessary for the preservation and wellbeing of society.

The purpose announced here is that society should continue to provide for the welfare and happiness of its members. The operative clause states that this purpose can be accomplished when government and subordination preserve the wellbeing of society.

In sum, this relationship between the prefatory "being necessary" absolute phrases and the clauses that follow them seems to me, in light of the range of applications for these constructions, to point to a late-18th-century habit of thought for many different kinds of problem-solvers and strategists in that period of U.S. history. Lawyers, engineers, generals, builders, politicians, and theologians all took advantage of this formulaic expression of a statement of purpose, problem, or principle followed by a statement of the means of achieving that purpose, solving that problem, or applying the principle. Seeing these 55 uses of that rhetorical and semantic device during the Federalist era allows us to see the text of the Second Amendment in a broader and richer context. The Second Amendment, too, observes the protocols of this structure: It states a purpose – in Justice Scalia's words, "to prevent elimination of the militia" – and then specifies the means of preserving these well-regulated militias, namely, the establishment of the right to bear arms.

Historians of the absolute phrase provide rich descriptions of it. Stump, Kortmann, Killie and Swann, Timofeeva, and others have thoroughly described, catalogued, and categorized the semantic relationships of an absolute phrase to its main clause; the absolute phrase is a truly versatile modifier, capable of providing very clear semantic signals or quite ambiguous ones. Absolute phrases can provide many different kinds of information, from cause, condition, concession, or time, to reason, manner, purpose, result, additional or accompanying circumstance, and exemplification or specification. Kortmann (1991: 173) refers to the "apparent tendency of *being*-adjuncts/absolutes [that is, absolute phrases that use the "being" verb form] to function as idiomatic constructions of causality." According to Timofeeva (2010: 31), such absolutes "establish a circumstantial framework within which the main predication holds"; Fonteyn and van de Pol (2015) observe that "sentence-initial adverbial clauses will express those adverbial relations that serve as frame-setting background information, i.e. temporal and causal relations." A quick reading of these 55 sentences with absolute phrases provides strong evidence that Justice Scalia accurately named the relationship between that absolute phrase and the main clause, the operative clause, that follows it – such a phrase is well described as announcing a purpose.

What we've observed in the "being necessary" absolute phrases from COFEA, especially in those instances that make a legal or normative case, is an instance of a special kind of relationship between some of these absolute phrases and the independent clauses that follow them. This kind of absolute phrase has as its central task the announcement of a purpose, and the main clause that follows the prefatory material provides the means for achieving that purpose. Grammatically, the main clause is still the main clause, but in these passages it is the absolute phrase that establishes a specific purpose; the main clause plays the operative or enacting role.

On this reading of the Second Amendment, the manifest interest in maintaining or enhancing the security of the state by providing for a functioning militia is the deepest purpose for advancing the amendment – it would be a very different amendment if the desire to maintain or enhance the security of the state were to be considered incidental to or possibly less important than some other unstated objective, such as hunting or self-defence, two other perfectly legitimate purposes, indeed, but neither of which appears in the phrase or the clause or is even remotely suggested by the language of the Second Amendment.

To see how this kind of absolute phrase worked in similar circumstances in the 18th century, consider the following passages from several legal cases and several legislative records of that period.

At this point, in the interest of demonstrating how many kinds of absolute phrase are engaged in this kind of writing, I have included extracts featuring absolute phrases of the "being necessary" sort but also using other adjectives than "necessary" in the S–LV–PA construction, the "It"-cleft versions with "necessary," and one passage that employs a noun in the predicate noun position (S–LV–PN).

A. Absolute phrases preceding legal or quasi-legal decisions (absolute phrases are in bold type; the incomplete verb and its complements are underlined):

(1) 1785: From Thomas Jefferson to John Adams, with Enclosure, 28 July 1785:

The parties, being desirous of promoting as much as possible the happiness of their citizens and subjects respectively and mutually, believing that a free and friendly intercourse between them will contribute much to this end, and that this intercourse cannot be established on a better footing than that of a mutual adoption by each of the citizens or subjects of the other, insomuch that while those of the one shall be travelling or sojourning with the other, they shall be considered to every intent and purpose as members of the nation where they are, entitled to all the protections, rights and advantages of it's native members, have, on mature deliberation, covenanted with each other, that the intercourse between all the subjects and citizens of the two parties shall be free and unrestrained:

(2) 1764: "The Quakers assisting, to preserve the lives of the Indians, in the barracks, vindicated":

and **this Accuser not being able to produce any Thing, more than calumnious Accusations, without advancing one single fact, whereby the Defendant hath broke any one Law, or even Transgress'd the Rules of Morality**, this Defendant is immediately discharged, and cleared of, and from, all the false seeming Accusations.

(3) 1794: Case law: "Between MARGARET FIELD, executrix of James Field, plaintiff, AND COLLIER HARRISON, and Christiana his wife, executrix of David Minge, defendants":

David Minge being dead and William Claiborne being insolvent, the creditors executrix, who could not maintain an action at common law against the representatives of the former, as is generally supposed, because the obligation being joint the right of action survived, brought a bill in equity, for recovering the money.

(4) 1764: Mr. Hutchinson, lieutenant-governor of the Massachusetts province:

The submission unto which proceedings of theirs being, as we apprehend, inconsistent with the maintenance of the laws and authority

here, so long enjoyed and orderly established under the warrant of his Majesty's royal charter, the upholding whereof **being** absolutely **necessary** for the peace and well being of his Majesty's good subjects here – This court doth therefore in his Majesty's name, and by his authority to us committed, by his royal charter, declare to all the people of this colony, that in observance of their duty to God and to his Majesty, and to the trust committed unto us by his Majesty's good subjects in this colony, we cannot consent unto or give our approbation of the proceedings of the abovesaid gentlemen, neither can it consist with our allegiance that we owe to his Majesty, to countenance any who shall in so high a manner go across to his Majesty's direct charge, or shall be their abettors or consentors thereunto.

(5) 1793: Case Law: Pennsylvania v. Bell, 1 Add. 156; Dec. 1793; Washington County Court:

The allegation of both being therefore necessary, if either be alledged on an impossible day, as the 31st of June, or if the murder be alledged to have been on a day different from the day on which the death is alledged to have been, though on the day on which the stroke was given, as if the jury find the stroke given on the 1st, and the death on the 10th of June, and so conclude that he murdered him on the 1st of June, when it appears from the indictment, that he was alive: this is repugnant and void:

B. Legislation: prefatory material with "whereas" clauses and an absolute phrase:

(1) 1777: Journals of the Continental Congress:

And whereas, in the session of this state, heretofore made, of territory for the government of the United States, the lines of which session could not be particularly designated, and **it being expedient and proper that the same should be recognised in the acts of this state**, Be It Enacted by the General Assembly of Maryland That all that part of the said Territory called Columbia, which lies within the limits of this state, shall be and the same is hereby acknow to congress, kc.

(2) 1777: Maryland state legislation:

174. Preamble. Whereas a number of citizens have claims against British merchants trading to this state before the revolution, and since peace those merchants have appointed factors or agents to collect the debts due to them from the citizens of this state, and those factors or agents never having notified, by public advertisement or otherwise, their power to adjust the debts of said merchants, those citizens who have claims against them are unable

to obtain a settlement of their accounts; and **it being necessary to secure as far as possible to our citizens their just debts**; therefore let it be enacted by the General Assembly of Maryland, That all such factors or agents, or their principals, now within this state, unless bond with security be given as herein after directly, shall, on or before the first day of August next, lodge with the auditor a list of all balances due to such merchants, upon oath; and any such factor, agent or merchant, who shall hereafter come into this state, shall, within four months from the time of coming into this state, lodge with the auditor a list upon oath of balances due to such merchants: and if they should neglect to deliver such list as aforesaid, then it may be lawful for the debtors of such merchants to plead the non-compliance with this law; and the several courts of justice within this state are hereby directed not to proceed, after the first day of August next, to give judgment against any citizen of this state, on any action brought by any of the said merchants, or their agents, as aforesaid, unless they produce a certificate from the auditor, certifying that this act has been complied with.

(3) 1798: Public Laws of the State of Rhode-Island and Providence Plantations Passed since the Session of the General Assembly in January, A.D. 1798:

Whereas the unexampled prosperity, unanimity and liberty, for the enjoyment of which this Nation is eminently distinguished among the nations of the earth, is to be ascribed, next to the blessing of God, to the general diffusion of knowledge and information among the people, whereby they have been enabled to discern their true interests, to distinguish truth from error, to place their confidence in the true friends of the country, and to defeat the falsehoods and misrepresentations of factious and crafty pretenders to patriotism; and **this General Assembly being desirous to secure the continuance of the blessings aforesaid, and moreover to contribute to the greater equality of the people, by the common and joint institution and education of the whole**: BE it enacted by the General Assembly, and by the authority established thereof it is hereby enacted, That each and every town in the State shall annually cause to be established and kept, at the expense of such town, one or more free Schools, for the instruction of all the white inhabitants of said town between the ages of six and twenty years, in reading, writing, and common arithmetic, who may stand in need of such instruction and apply therefor.

(4) State constitutions: prefatory material with absolute phrase and no "whereas" clauses:

1774: Journals of the Continental Congress; official records (#9, below):

Religion, Morality and knowledge being necessary to good government and the happiness of mankind, Schools and the means of education shall forever be encouraged.

1781: CHAP. V. SECT. II. of the Constitution of the Province of Massachusetts Bay:

WISDOM and knowledge, as well as virtue, diffused generally among the body of the people, being necessary for the preservation of their rights and liberties; and as these depend on spreading the opportunities and advantages of education, in the various parts of the country, and among the different orders of the people, it shall be the duty of the legislatures and magistrates, in all future periods of this commonwealth, to cherish the interests of literature and the sciences, and all seminaries of them;

(5) Executive order from the president:

1793: To Alexander Hamilton from George Washington [8 August 1793]:

The present being, & being likely to continue for some time a favorable season for purchases of the Public Debt, and as it appears that the whole, or the greatest part of the unexpended monies of the foreign loans heretofore made, will be requisite for satisfying the approaching installments of our Debt to France, which it has been judged expedient to pay without deduction for any prior advance. **It appearing moreover from the statements submitted by you to my consideration, that the sum authorised to be borrowed by the Act intitled "an Act making provision for the reduction of the public debt," there may still be procured for the purpose of that act the sum of one million five hundred & fifteen thousand & ninety eight Dollars & Eleven cents**: I do therefore direct & require that you take immediate measures for obtaining a Loan or Loans to the extent of the sum aforesaid, to be applied to purchases of the public debt pursuant to the provisions of the act above mentioned.

(6) Advice or counsel to the President:

1783: To George Washington from Timothy Pickering, 22 April 1783:

But **the military establishments therein proposed being inadequate to the general defence of the country (which doubtless must engage your attention as well as that of Congress**) I have thought it would coincide with your views to lay before you my sentiments respecting a militia; which are therefore subjoined.

10 The Prefatory-materials canon

In *Heller*, a discussion of the grammar and semantics of the Second Amendment should provide a very obvious opportunity to apply originalist practices to a reading of the amendment. In their 2012 book *Reading law: the interpretation of legal texts*, Scalia and Garner describe 37 canons of legal interpretation, that is, "principles for interpreting all types of legal instruments, from constitutions to statutes to ordinances to regulations to contracts to wills" (2012: 51). The Prefatory-Materials Canon is number 34 in that list (2012: 217–220).

The Prefatory-Materials Canon is invoked at exactly this point of *Heller* in order to assert that the prefatory material, the absolute phrase, cannot be used to "limit or expand the scope of the operative clause." Justice Scalia's application of this canon at this point in the argument is not improper, but it is deeply counter-intuitive. The common-sense reading of the amendment aligns very well with the reading of many similar passages from COFEA. Separating the prefatory material from the operative clause creates a very different Second Amendment, one with an announcement of purpose that is declared irrelevant. This new Second Amendment consists of nothing more than an operative clause, an operative clause that has no announced purpose to which it is held accountable.

The Prefatory-Materials Canon is referred to in the text of *Heller* itself and in the notes to the opinion that provide bibliographic and other supporting information. The simplest explanation and implementation of this canon in *Heller* is as follows:

> The Second Amendment is naturally divided into two parts: its prefatory clause and its operative clause. <u>The former does not limit the latter grammatically, but rather announces a purpose.</u> (*Heller* 2008: 3; my underscore)

Likewise, on page 4 of *Heller* there is this sequence of sentences (I've omitted the illustrative materials that appear between these sentences):

> Logic demands that there be a link between the stated purpose and the command. [...] That requirement of logical connection may cause a prefatory clause to resolve an ambiguity in the operative clause. [...] But apart from that clarifying function, a prefatory clause does not limit or expand the scope of the operative clause. See F. Dwarris, A General Treatise on Statutes 268–269 (P. Potter ed. 1871) (hereinafter Dwarris; T. Sedgwick, The Interpretation and Construction of Statutory and Constitutional Law 42–45.
> (2d ed. 1874)

Note 3 includes this further comment on and summary by Sutherland:

> As Sutherland explains, the key 18th-century English case on the effect of preambles, Copeman v. Gallant, 1 P. Wms. 314, 24 Eng. Rep. 404 (1716), stated that "the preamble could not be used to restrict the effect of the words of the purview." J. Sutherland, Statutes and Statutory Construction, 47.04 (N. Singer ed. 5th ed. 1992). This rule was modified in England in an 1826 case to give more importance to the preamble, but in America "the settled principle of law is that the preamble cannot control the enacting part of the statute in cases where the enacting part is expressed in clear, unambiguous terms." *Ibid.*

One of the notes referring to this canon leads to *Emerson* (2001), which has in it a review of the Second Amendment and the collective-right versus the individual-right traditions; there is a substantial discussion of the status of prefatory materials in determining the meaning and the scope of an operative clause. *Emerson* provides in note 32 a history of this canon from sources in the late 19th century. These explanations of the history of the canon are reinforced by the 2012 book written by Antonin Scalia and Bryan Garner, *Reading law: The interpretation of legal texts*. In Chapter 34 (pp. 217–220), "Prefatory-Materials Canon: A preamble, purpose clause, or recital is a permissible indicator of meaning," Scalia and Garner consider the uses and the limitations of invoking the prefatory material, but one condition stands out: If the operative clause is unclear or ambiguous, it is necessary to look to the prefatory material for assistance in determining the meaning of the ambiguous word or phrase in the operative clause. If the disagreement between *Heller* and Baron on the meaning of "to bear arms" makes the operative clause unclear or ambiguous, the prefatory absolute phrase must be consulted, and if it is consulted, the scope of the operative clause would seem to be restricted to the arming of persons who serve in a militia.

11 Conclusion

In these two instances – the meaning of "to bear arms" and the meaning and scope of the introductory absolute phrase – the two corpora I have referred to seem to be very relevant for the discussion of the meaning of the Second Amendment. The *Heller* position on these two elements is put into question by the words, phrases, and clauses that COFEA has made available from the very era in which the Second Amendment was written and ratified. The absolute phrase that announces the purpose of the Second Amendment retains its function of being "a permissible indicator of meaning," and the common-sense reading of the amendment seems to support the older, traditional interpretation of this amendment, namely, that the bearing of arms referred to in the operative clause is done in support of well-regulated militias, not for hunting and self-defence.

References

Corpora

Corpus of Early Modern English; 1,107,365,393 words, 1475–1800; BYU Law (2018). https://lawcorpus.byu.edu/

Corpus of Founding Era American English; 136,848,583 words, 1760–1799; BYU Law (2018). https://lawcorpus.byu.edu/

Works cited

Baron, Dennis. 2018. Antonin Scalia was wrong about the meaning of 'bear arms.' *Washington Post*. 21 May 2018. https://www.washingtonpost.com/opinions/antonin-scalia-was-wrong-about-the-meaning-of-bear-arms/2018/05/21/9243ac66-5d11-11e8-b2b8-08a538d9dbd6_story.html

District of Columbia, et al., Petitioners v. Dick Anthony Heller. 2008. http://www.supremecourt.gov/opinions/07pdf/07-290.pdf

Fonteyn, Lauren & Nikki Van De Pol. 2016. Divide and conquer: The formation and functional dynamics of the Modern English ing-clause network. *English Language and Linguistics* 20 (2). 185–219.

Killie, Kristin & Toril Swan. 2009. The grammaticalization and subjectification of adverbial–ing clauses (converb clauses) in English. *English Language and Linguistics* 1 (3). 337–363.

Kortmann, Bernd. 1991. *Free absolutes and adjuncts in English: Problems of control and interpretation*. London & New York: Routledge.

Scalia, Antonin & Bryan A. Garner. 2012. *Reading law: The interpretation of legal texts*. St. Paul, MN: West Group.

Stump, Gregory T. 1988. *The semantic variability of absolute constructions*. Dordrecht, Boston & Lancaster: Springer.

Timofeeva, Olga. 2010. *Non-finite constructions in Old English with special reference to syntactic borrowing from Latin*. Helsinki: Société Néophilologique.

United States of America, Plaintiff-appellant, v. Timothy Joe Emerson, Defendant-appellee, 270 F.3d 203 (5th Cir. 2001). https://law.justia.com/cases/federal/appellate-courts/F3/270/203/545404/

Appendix I

55 sentences from COFEA with S–V–PA absolute phrase in the initial position (in chronological order)

1. 1764. Thomas Hutchinson, lieutenant-governor of the Province of Massachusetts Bay. Court ruling:

The submission unto which proceedings of theirs being, as we apprehend, inconsistent with the maintenance of the laws and authority here, so long enjoyed and orderly established under the warrant of his Majesty's royal charter, **the upholding whereof being absolutely necessary for the peace and well being of his Majesty's good subjects here** – This court doth therefore in his Majesty's name, and by his authority to us committed, by his royal charter, declare to all the people of this colony, that in observance of their duty to God and to his Majesty, and to the trust committed unto us by his Majesty's good subjects in this colony, we cannot consent unto or give our approbation of the proceedings of the abovesaid gentlemen, neither can it consist with our allegiance that we owe to his Majesty, to countenance any who shall in so high a manner go across to his Majesty's direct charge, or shall be their abettors or consentors thereunto.

2. 1765. Samuel Hopkins A.M., Minister of the Gospel in Great Barrington. Religious writing:

This knowledge being absolutely necessary in order to salvation, he who is without it, has a fatal bar in the way of his salvation:

3. 1770. James Dana, D.D., Pastor of the First Church in Wallingford. Religious writing:

For **the acts of the will, not being necessary in their own nature, but by connection with a cause that is so;** and no cause being thus necessary but he who existed from eternity, it undeniably follows, that every sinful volition proceeds ultimately from him, as the cause and source.

4. 1770. James Dana, D.D., Pastor of the First Church in Wallingford. Religious writing:

For the acts of the will, not being necessary in their own nature, but by connection with a cause that is so; **and no cause being thus necessary but he who existed from eternity**, it undeniably follows, that every sinful volition proceeds ultimately from him, as the cause and source.

5. 1770. William Robertson, D.D., historiographer to His Majesty for Scotland. History:

His honour, however, and passions concurred in preventing him from relinquishing his scheme of a divorce, which he determined to accomplish by other means, and at any rate; and **the continuance of Francis's friendship being necessary to counterbalance the Emperor's power**, he, in order to secure that, not only offered no remonstrances against the total neglect of their allies, in the treaty of Cambray, but made Francis the present of a large sum, as a brotherly contribution towards the payment of the ransom for his sons.

6. 1771. Dr. Samuel-Auguste Tissot. *Avis au peuple* [*Medicine and health*]. Trans from the French by James Kirkpatrick. Medical writing:

    Persons of Education must find a Pleasure, I conceive, in making People understand on these Occasions, which are so frequent, that **the Air being more indispensably necessary to us, if possible, than Water is to a Fish**, our Health must immediately suffer, whenever that ceases to be pure;

7. 1773. James Dana, D.D., Pastor of the First Church in Wallingford. Religious writing:

    For though by their hardness and impenitency they treasure up to themselves wrath, yet **their incorrigibleness in sin, and their eternal punishment for those vices to which they were determined by providence, being necessary to advance the holiness and happiness of the universal system,** they may be excited and exhorted to persist in sin by the motive of public benevolence; and may have this consolation under the apprehension of that indignation and wrath, tribulation and anguish, which await them, that their destruction soul and body in hell, is for the good of the universe!

8. 1774. *Journals of the Continental Congress*. Official records:

    But that **the articles contracted for being necessary for the Army**, and the intentions of the Memorialists in making the contract being apparently directed to promote the public interest, and moreover the terms not being unfavorable, it ought to be complied with as far as possible.

9. 1774. *Journals of the Continental Congress*. Official records:

    Religion, Morality and knowledge being necessary to good government and the happiness of mankind, Schools and the means of education shall forever be encouraged.

10. 1774. *Journals of the Continental Congress*. Official records:

    the accounts exhibited to Congress in the printed book accompanying the Letter referred to your Committee do not appear to have been examined and adjusted, since the resignation of the late Superintendent, by any person duly authorized for that purpose, and **this measure being previously necessary, in the opinion of your Committee to enable Congress to determine, with**

propriety on the Subject of said Letter, they beg leave to recommend the following resolution

11. 1775. The Committee of Safety's account of the battle of Bunker Hill, 25 July 1775. History:

    Accordingly on the 16th. Ultimo Orders were issued that a Detachment of one thousand Men should march that Evening to Charlestown, and entrench upon that Hill; just before 9 oClock they left Cambridge, and proceeded to Breeds Hill, situated on the further Part of the Peninsula West of Boston, for by some Mistake this Hill was marked out for the Entrenchment instead of the other; **many things being necessary to be done preparatory to the Entrenchments being thrown up, which could not be done before lest the Enemy should discover and defeat the Design**, it was nearly 12 oClock before the Works were entered upon.

12. 1776. William Northcote, surgeon. Medical writing:

    nevertheless, **salivation not being necessary to the cure**, the use of the medicine No. 66. is to be left off immediately on the first signs of a spitting coming on.

13. 1777. American Commissioners: Benjamin Franklin, Silas Deane, and Arthur Lee; Memorandum for Vergennes, 1 March 1777. Memorandum:

    A Port in Europe being Necessary for the Shipps of War of the United States, and the Treaty Subsisting between France and Great Brittain not permitting one to be granted in this Kingdom; it is hoped that it will not be disagreeable That Application should be made elsewhere, and on this Subject intreat his Excellency's advice and direction which on this and every other Concern will be attended to by them in the closest manner, as well as be received with the utmost respect and Gratitude.

14. 1777. Brigadier General John Cadwalader. Plan for Attacking Philadelphia, 24 November 1777. Official records:

    As the Enemy have made very considerable Detachments from their main Body to New Jersey under the Command of Lord Cornwallis; and **a considerable number of men being necessary to defend the several Posts on the Islands which are at least 7 miles from the Lines** it may be very proper to consider whether a successfull attack cannot be made on the City.

15. 1779. From George Washington to James Wilkinson, 6 December 1779. Official correspondence:

    Your presence here being absolutely necessary you will be pleased to come on yourself and leave the management of the removal to your Assistant.

16. 1780. William Livingston. Letter to George Washington, 4 August 1780. Official correspondence:

    Dear Sir I received your Excellency's Favour of the first Instant yesterday, & **the advice of Council being necessary to authorize me to comply with your Requisitions**, I lost no Time in summoning one for that purpose –

17. 1780. George Muter. Letter to Thomas Jefferson, 1 September 1780. Official correspondence:

    A new regulation respecting the pay of wagonmasters, foragemasters, & c., being absolutely necessary, a memorandum on that subject from Mr. Rose is enclosed for TJ's consideration.

18. 1781. George Washington. Letter to Board of War, 12 January 1781. Official correspondence:

    He expects to go to France and **dispatch being necessary to his plan**, I have thought proper to refer him thus immediately to the Board.

19. 1781. George Muter. Letter to Thomas Jefferson, 20 January 1781. Official correspondence:

    An estimate being necessary of the sum these articles will cost before we can judge whether it can be furnished, we shall be glad of such an estimate.

20. 1781. Marquis de Lafayette. Letter to Thomas Jefferson, 16 March 1781. Official correspondence:

    Harmony being necessary to the public wellfare, I thought it was Better for me Not to Make Any Mention of this Affair, But Am no less obliged to the Confidence which Has Been put in Me, and which I am warmly desirous to deserve.

21. 1781. Thomas Jefferson. Letter to the Commander of the Essex County Militia, 8 May 1781. Official correspondence:

    Cavalry in a due proportion being as necessary as Infantry you will be pleased to permit and even to encourage one tenth Part of those who are to come into Duty as above required to mount and equip themselves as Cavalry.

22. 1781. Chap. v. Sect. ii. of the Constitution of the Province of Massachusetts Bay. Official records:

   WISDOM and knowledge, as well as virtue, diffused generally among the body of the people, being necessary for the preservation of their rights and liberties; and as these depend on spreading the opportunities and advantages of education, in the various parts of the country, and among the different orders of the people, it shall be the duty of the legislatures and magistrates, in all future periods of this commonwealth, to cherish the interests of literature and the sciences, and all seminaries of them [. . .]

23. 1785. Benjamin Franklin & Thomas Jefferson, American Commissioners. Letters to Baron de Thulemeier, 26 May 1785. Official correspondence:

   But **the signature of at least two of our number being necessary**, and Mr. Adams being called away by his mission to the court of Great Britain, and another of us rendered unable by age and a painful malady to perform a land journey, there is a difficulty in meeting with your Excellency for the purpose, either at any intermediate place, or at that of your residence, (which in respect to the King we might otherwise willingly do).

24. 1787. Governor Robert Morris. *The Records of the Federal Convention of 1787*. Official records:

   Some check being necessary on the Legislature, the question is in what hands it should be lodged.

25. 1790. Alexander Hamilton. Treasury Department Circular to the Commissioners of Loans, 1 November 1790. Memorandum:

   This being necessary, I am to request that if any have gone from your Office without your name, you will apply for them to the holders and add your signature.

26. 1790. Annals of the Congress of the United States 1st Congress to 18th Congress, 1st Session (1789–1924). Official records:

   Suppose one mechanic art to be heavily taxed and others not taxed at all. This, at first, will be oppressive; but **the art being necessary in society**, other arts must contribute to support it under the increased burden.

27. 1790. Annals of the Congress of the United States 1st Congress to 18th Congress, 1st Session (1789–1924). Official records:

The Committee had not much time to consider the subject; but **something being necessary to be done,** they had agreed to make the report they had made.

28. 1790. Richard Baxter. *The saints everlasting rest: or, A treatise of the blessed state of the saints in their enjoyment of God in glory.* Religious writing:

    This being so necessary an attendant of love, and being excited much by the same considerations, I suppose you need the less direction.

29. 1790. Annals of the Congress of the United States 1st Congress to 18th Congress, 1st Session (1789–1924). Official records:

    Foreign coin being therefore necessary to pay the commercial balance due from us to foreign countries, it would certainly be imprudent to call it in.

30. 1791. Pierre-Charles L'Enfant. Letter to George Washington from, 19 August 1791. Official correspondence:

    from these consideration **a better security of funds being necessary to combine a plan of operation the good of which can only be Insured aided by punctual payements and regular and plentifull supply of Materials** it will expedient first to devise the necessary means considering that Economie in a poursuit of this nature lay in being aided with numerous hands with a power of pouring means there were they may accelerat the leveling of difficulties frequent to enconter.

31. 1792. Joel Barlow. *Advice to the privileged orders, in the several states of Europe, Part I.* Political writing:

    The same may be said of all regulations that arise from the social compact. It is a truth, I believe, not to be called in question, that every man is born with an imprescriptible claim to a portion of the elements; which portion is termed his birth-right. Society may vary this right, as to its form, but never can destroy it in substance. She has no controul over the man, till he is born; and **the right being born with him, and being necessary to his existence,** she can no more annihilate the one than the other, though she has the power of new-modeling both.

32. 1792. Thomas Jefferson. Memorandum of Conference with the President on Treaty with Algiers, 11 March 1792. Memorandum:

    The subseqt. approbation of the Senate being necessary to validate a treaty they expect to be consulted before hand if the case admits.

33. 1792. Thomas Jefferson. Memorandum of Conference with the President on Treaty with Algiers, 11 March 1792. Memorandum

   So **the subseqt. act of the Repr. being necessary where money is given**, why should not they expect to be consulted in like manner where the case admits?

34. 1793. George Washington. Letter to Anthony Whitting, 19 May 1793. Official correspondence:

   A Stirring, lively & spirited man, who will act steadily & firmly, being necessary, I authorise you to get one if you should part with Butler

35. 1793. Case Law: Pennsylvania v. Bell, 1 Add. 156; Dec. 1793; Washington County Court. Legal writing:

   The allegation of both being therefore necessary, if either be alledged on an impossible day, as the 31st of June, or if the murder be alledged to have been on a day different from the day on which the death is alledged to have been, though on the day on which the stroke was given, as if the jury find the stroke given on the 1st, and the death on the 10th of June, and so conclude that he murdered him on the 1st of June, when it appears from the indictment, that he was alive: this is repugnant and void

36. 1794. Defence of the Constitutions of Government of the United States of America. Political writing:

   The people had three privileges; to choose magistrates (yet all the great employments must be confined to the patricians); to enact law; and to determine concerning war, when proposed by the king: but **the concurrence of the senate being necessary to give a sanction to their decisions**, their power was not without controul.

37. 1794. Thomas Sim Lee. Letter to Alexander Hamilton from Thomas Sim Lee, 28 August 1794. Official correspondence:

   The Consent of the Legislature of the United States being necessary to the imposition and collection of the Tonnage duty contemplated by this Act as the means of its Execution, a Law was passed at the last Session of Congress by which the Operation of the above-mentioned Act of Assembly is assented to "so far as to enable this State to collect a duty of one cent per Ton on all Vessels coming into the District of Baltimore from a Foreign Voyage for the purpose in the said Act intended."

38. 1794. Minutes of the proceedings of the committee, appointed on the 14th September, 1793, by the citizens of Philadelphia. Committee minutes:

A large supply of sheets, blankets, & c. being immediately necessary, Thomas Wistar and Henry Deforest are requested to take their instructions from the managers and procure every necessary that may be required for the use of the sick at the Hospital.

39. 1795. Hugh H. Brackenridge. *Incidents of the insurrection in the western parts of Pennsylvania, in the year 1794.* History:

I do not mean to question the necessity, for nothing else could justify it, of making the arrest in the night, and by squadrons of horse; but I only take notice, that, **this being necessary**, it was the greater hardship for a man to be arrested, who was a good citizen; for instead of being treated, as far as consistent with confinement, with all the delicacy that a still existing presumption of innocence demands, and which the mild habits of an arrest, by a common civil officer, gives, he is subjected to the insults, which may naturally be expected from those who, having just before thought of fighting and killing, are disposed now to have, at least the satisfaction of cursing, or starving, or otherwise abusing the people.

40. 1795. Timothy Pickering. Letter to George Washington, 17 September 1795. Official correspondence:

I thought it proper however to lay it before you: at the same time it appears to be so clear a case, that I have written an answer to Mr Paleski, suggesting that the prolonging of a treaty is tantamount to the making of a treaty, in which **the act of the Senate being necessary,** the object of his letter must of course be postponed till the Senate assembles.

41. 1795. Oliver Evans. *The young mill-wright & miller's guide.* Construction advice:

And again, if the fire be large, and the chimney too small, smoke cannot be all vented by it, **more air being necessary to supply the fire than can vent up the chimney**, it must spread in the room again, which, after passing through the fire and being burnt is suffocating.

42. 1796. Emerich de Vattel. *Law of Nations: or, Principles of the Law of Nature; Applied to the Conduct and Affairs of Nations and Sovereigns*; Trans. Anon from French. Legal writing:

> A declaration of war being necessary, as a further effort to terminate the difference without the effusion of blood, by making use of the principle of fear, in order to bring the enemy to more equitable sentiments, – it ought at the same time that it announces our settled resolution of making war, to set forth the causes of that resolution.

43. 1796. Captain James Cook & Captain James King. *A voyage to the Pacific Ocean*. History:

> But **a supply of water being necessary before he could execute that design,** he determined, with a view of procuring this essential article, to search the coast of America for a harbour, by proceeding along it to the southward.

44. 1796. Emerich de Vattel. *Law of Nations: or, Principles of the Law of Nature; Applied to the Conduct and Affairs of Nations and Sovereigns*; Trans. Anon from French. Legal writing:

> **Usucaption and prescription being so necessary to the tranquillity and happiness of human society,** it is justly presumed that all nations have consented to admit the use of them as lawful and reasonable

45. 1797. *Rules of discipline and Christian advices of the Yearly Meeting of Friends for Pennsylvania and New Jersey, first held at Burlington in the year 1681, and from 1685 to 1760, inclusive, alternately in Burlington and Philadelphia: and since at Philadelphia*. Quaker ordinances:

> **Some explanation respecting the authority of Monthly Meetings to disown our Youth or others who depart from that simplicity which Truth requires, and who run into, * and copy after the vain and extravagant fashions of the world in their dress and address, being necessary, and coming under solid and weighty consideration,** it appears to be the sense of this Meeting, that if after patient Labour in the Spirit of meekness and wisdom, such cannot be reclaimed, Monthly Meetings may give forth Testimonies of Disownment against them.

46. 1797. Anon. *Female friendship, or The innocent sufferer: A moral novel*. Prose fiction:

> Charles having no longer the same reasons for absenting himself as heretofore, readily coincided with his proposal: but **some time being necessary, for him to get rid of his business,** our hero was detained longer at Paris, than he at first intended.

47. 1797. Jonathan Edwards. *A dissertation concerning liberty and necessity*. Religious writing:

and also because there being no motion without a particular determination one certain way, and **no one determination being more necessary than another**, an essential and necessary tendency to motion in all determinations equally, could never have produced any motion at all.

48. 1797. Jonathan Edwards. *A dissertation concerning liberty and necessity*. Religious writing:

There being no motion, i.e. volition, without a particular determination one certain way, and **no one determination being in nature more necessary than another**, an essential and necessary tendency to volition in all determinations equally, could never have produced any volition at all.

49. 1797. James Butler. *Fortune's foot-ball: or, the adventures of Mercutio*. Prose fiction:

Speedy and effectual measures being absolutely necessary, they devoted one whole night in deliberating what was most proper to be done; and after proposing and rejecting several plans, determined, that Mercutio should immediately provide for their transportation to England, if possible; if not, to some other part of the world.

50. 1797. James Butler. *Fortune's foot-ball: or, the adventures of Mercutio*. Prose fiction:

Though this was an unwelcome piece of intelligence, yet **a compliance being absolutely necessary,** they parted, with reluctance, Leonora having previously pressed a very valuable ring on the finger of Mercutio, which she requested him to wear in testimony of her unalterable attachment.

51. 1798. James McHenry. Letter to George Washington, 29 November 1798. Official correspondence:

A settlement of his accounts, for the recruiting service, being necessary, a previous muster of the recruits was directed, when several of the men were mustered out. and the accountant of this Department by letter to the Secretary of War dated the 3rd of July last reported alligations of irregularities in the conduct of Captain Miller, and referred to him to determine whether the bounty paid, and charged by Captain Miller to recruits, now reported to have been unfit for service at the time of inlistment, should not be deducted from his account.

52. 1798. Marie-Jeanne Roland. An *appeal to impartial posterity*. Trans. from the French. History:

An excursion to the country being necessary for the perfect re-establishment of my health, we went to breathe its salutary air at the house of M. and Madam Besnard, with whom two years before my mother and I had spent almost the whole month of September.

53. 1798. Benjamin Count of Rumford. *Essays, political, economical, and philosophical.* Trans. from German. Political writings:

A knowledge of the cause of the ascent of Smoke being indispensably necessary to those who engage in the improvement of Fire-places, or who are desirous of forming just ideas relative to the operations of fire, and the management of heat, I shall devote a few pages to the investigation of that curious and interesting subject.

54. 1799. Samuel Fiske. An oration pronounced at Claremont, on the anniversary of American independence. July 4th, 1799. Patriotic oration:

SOCIETY being necessary for the welfare and happiness of man; so government and subordination are necessary for the preservation and well being of society.

55. 1799. Alexander Hamilton. Letter to James Miller, 22 October 1799. Official correspondence:

The axes, Cross cut saws, files and Grindstones, being necessary to begin the work they must not be delayed for the purpose of procuring the others.

Jacob Livingston Slosser
Experimental legal linguistics: A research agenda

Introduction

Linguistics is but one entry into the interdisciplinary endeavour that makes up the study of legal thought. Whether it is the work of economics, psychology, sociology, or linguistics, the legal academy can be characterised as being quite comfortable with studying itself through the lenses of other disciplines (see Solan & Tiersma 2012). Much of the interdisciplinary involvement between law and linguistics has been in the form of forensic linguistics-like efforts, using methods adopted from linguistics to help solve legal problems such as: authorship identification (Cotterill 2010), determining "ordinary meaning" across a corpus of law (Solan 2020), or issues with multilingual institutions (McAuliffe 2020), to name only a few examples. In much of this work, the utility of linguistic methods is that they aid in understanding the law's plethora of abstractions regarding legal norms, principles, rules, or standards. Linguists have important contributions to make in understanding legal meaning – perhaps their most fundamental role with respect to the law.[1] As Friedmann Vogel has pointed out, "Norms have to be 'performed' by concrete actors, and have to be construed actively by working with and on statutory (or preceding judicial) texts" (2018: 1342). Notwithstanding the factors and influences affecting institutions, praxis, rational choice, etc., at its foundation this performance is the language that reflects the cognitive action

---

[1] This approach is not exclusive to law and language studies, and has traditionally theorised norm-building through disciplinary lenses such as the sociology of law, with a focus on a habitus and the cognitive elements of legal practice (DiMaggio 1997; Ignatow 2007; Slosser & Madsen, forthcoming), the economics of law with a focus on rational actor theory (see van Aaken & Broude 2019), or the diverse lenses of legal philosophy (see e.g. Dunoff & Pollack 2017; Bianchi 2016), among various others.

Acknowledgements: Jacob Livingston Slosser is a Carlsberg Postdoctoral Research Fellow, in the University of Copenhagen Faculty of Law: Danish National Research Foundation's Centre of Excellence for International Courts (iCourts). Contact: jacob.slosser@jur.ku.dk. https://orcid.org/0000-0003-4913-4790. This work was produced with the support of the Carlsberg Foundation

Jacob Livingston Slosser, University of Copenhagen

https://doi.org/10.1515/9783110720969-006

of its performers. The present chapter focuses on one strain of cognitive theory, Conceptual Metaphor Theory, for one type of performance: reasoning from prior cases to solve a legal problem. It focuses on a missing counterpart to other legal linguistic approaches: experimental falsification and validation. This research agenda outlines the promise of the cognitive linguistic experimental paradigm to counter the distinction hypothesis of legal meaning,[2] strives to discover the generalizable across legal language, and aims to capture nuance through an adjustable and reproducible experimental framework. The experimental framework is designed to highlight and test assumptions about the force of underlying mechanisms that affect meaning, not only within within legal thought but within thought itself.

To introduce this research agenda, I must start with a few necessary hedges. This chapter will focus on one subset among a variety of approaches one could use to develop an experimental framework: cognitive linguistics, and, in particular, Conceptual Metaphor Theory. This reduction in scope highlights the kinds of contributions and experiments that might be done within a single framework. The goal here is twofold. One, to demonstrate the strength of Conceptual Metaphor Theory for exploring how a change in language may cause a change in thought. Two, to demonstrate a program that can serve as a foundation for not just single experiments and findings, but reproducible results that can build a body of evidence. In this way, linguistic research in law can merge with some of the literature on legal decision-making, by testing assumptions about legal cognition as a separate entity. Among the assumptions tested include questions such as: does expertise matter to legal reasoning? Are there significant differences in reasoning across languages, legal cultures, and systems and time frames? Some empirical work has been undertaken by analysing across corpora, by case study, or through detailed discourse analysis (Winter 2001; Slosser 2019), but many questions remain as to how reliable or comparable a written judgment may be as a proxy for legal thought, and whether certain constructions are used as genuine artefacts of reasoning or as post-hoc justifications. Through adjustable and reproducible experiments like the model laid out in this chapter, these kinds of questions can be approached in a more scientifically falsifiable way by independent researchers.

Section one briefly reviews the basics of Conceptual Metaphor Theory, and lays out the evidence for its role in reasoning from the literature. It explains what kinds of questions we can ask using an experimental model. Section two establishes the theoretical basis for how Conceptual Metaphor Theory can apply to the

---

2 See generally the distinctness hypothesis of legal language, e.g. Tiersma (1999).

traditional ideas of reasoning from prior cases,[3] analogical reasoning. Section three presents a basic, adaptable example of the kind of experiment this article proposes. This model will serve as a prototype, demonstrating how, through adjustments to participant selection, environmental design, legal area covered, and other factors, a robust body of evidence for the force of language on legal thought can be produced. Finally, the chapter concludes by exploring some of these experimental controls, and what might be learned by transferring research done in social cognitive science and psycholinguistics to legal thought.

1 The role of Conceptual Metaphor Theory in reasoning

It is beyond the scope of this chapter to give an exhaustive overview of the field of cognitive linguistics (for such an overview, see e.g. Croft & Cruse 2004; Evans & Green 2006). Instead, this section focuses the underlying theoretical commitments of cognitive linguistics, and on the major units of analysis that are applied to legal meaning in this chapter: *conceptual metaphor* and *frames*. The premise underlying Conceptual Metaphor Theory is that metaphor, although often misunderstood to be pure flourish and rhetoric, is basic to thought itself (Lakoff & Johnson 2008), though to what exact extent this is true remains a point of debate in the field (see Pinker 2007). While some metaphor can instantly be seen for its rhetorical effects, it is the more hidden and implicit presence of metaphor in language that offers much for the legal linguist to consider. From the perspective of Conceptual Metaphor Theory, metaphor is a cognitive mechanism that structures thought in co-relational mappings that people use to derive abstractions from concrete experience and from motor-sensory input. That is to say, when a hearer understands a phrase that is common to both legal and non-legal corpora like "the reasons *carried* more *weight*," it is through the hearer's prior experience of weighted things in concrete reality that the metaphorical sense of a non-concrete thing (a reason, a thought) having weight is understood. Following from the definition found in Thibodeau et al. (2007), there are three main components in a conceptual metaphor: "a source domain, a target (or topic) domain, and a mapping between them" (2017: 853). In the above example the source domain is *weight*,

---

3 I use the phrase "reasoning from prior cases" here rather than strictly "precedent" to highlight the applicability of this experimental approach for reasoning in both common and in civil law systems. Adjustments could be made to accommodate for the different styles of a similar effort. See generally Komárek (2013).

and the target is *importance* – things that are important are *heavy*, or *carry weight*. The mapping between the concrete and the abstract allows the hearer to infer further that *lighter* reasons are not as good, or lack *gravity*, or to *balance* considerations against one another. Extending the concept further, we might avoid a certain topic because it is *too heavy*, or we dismiss a fleeting thought as it *floats on by*. This internal physics – see here Talmy's concept of *force dynamics* (Talmy 2000) – is not just a rhetorical choice agreed upon tacitly by both speaker and hearer, it is a function of how we think. This conceptual metaphor is formalised as IMPORTANCE IS HEAVY, using the standard TARGET IS SOURCE formula for conceptual metaphors, where the concrete *source* experience (HEAVY WEIGHT) maps onto the abstract *target* concept (IMPORTANCE).

There are countless examples of these types of mappings in speech: LOVE IS A JOURNEY ("we've *come to* a *crossroads*"); GOOD IS UP ("she received a *high* mark on the exam"); ARGUMENT IS WAR ("he couldn't *defend* his argument"); or perhaps, LAW IS A STRUCTURE ("the case law is *founded upon*; [. . .] is *settled*"), among many others (Lakoff 2008; Evans 2007; Slosser 2019).[4] Source domains are most often concrete concepts relating to time, space, weight, arrangement, substance; bodily experiences that allow the for the porting of those experiences onto the abstract categories we build, both individually and culturally. What makes conceptual metaphor so powerful is the coherent systematicity of metaphor that, "Concepts are not simply 'one offs,' they fall into a systematic frame that helps categorise and give meaning to abstract concepts through a reliance on *embodied experience*: abstract concepts are based in the more tactile realm and gained through bodily experiences" (Slosser 2019: 597). These frames organise the mappings between concrete and abstract so that meaning can be inferred where "linguistic units serve as prompts for an array of conceptual operations and the recruitment of background knowledge" (Evans & Green 2006: 45). This understanding of meaning-making posits that metaphor has a significant effect on determining meaning, via how metaphor helps build concepts in language.

In many experimental settings – within the field of cognitive linguistics, psycholinguistics, social psychology, political communication studies, among others – metaphoric frames have been used to cue a participant to think of one problem in terms of another. There are a few terms with which one might describe the systemisation of structural mapping, and the level of conceptualization at which they act: *schemas, scenarios, scene, gestalt, frames, cognitive models, idealised cognitive models, mental spaces* – and a number of others (for disambiguation see

---

4 Collections of metaphorical frames can be found at UC Berkley's MetaNet wiki repository: https://metaphor.icsi.berkeley.edu/pub/en/index.php/MetaNet_Metaphor_Wiki

Kovecses 2017). These are not interchangeable, but for sake of simplicity this chapter focuses on the term *frame*, defined as a device that relies on a systematic relationship of attributes used to evoke certain aspects of similarity between target and source domains.[5] To use the above example, if, in controlled experiments, one wanted to invoke similarities between heaviness and importance, one would use the frame WEIGHT in descriptions of, for example, arguments, reasons, or considerations.[6] In fact, one might not even use linguistic descriptions to cue the weight frame in the research participants but could cue participants by providing them with heavier clipboards on which to fill out surveys. In one such experiment, it was found that participants with heavier clipboards rated issues as more important than did their counterparts who were given lighter clipboards (see Jostmann et al. 2009). Although they act by the same mechanism, there is a distinction here between the type of cueing of frames that are done implicitly or explicitly.[7] Those studies that involve implicit cueing are those that use "very subtle or implicit cues" (Keefer & Landau 2016: 399), like the study above cueing the conceptual metaphor of IMPORTANCE IS WEIGHT ("some topics are heavy; some considerations carry more weight than others; and so on" [Keefer & Landau 2016: 399]) by using the weight of a clipboard held by the participants. Explicit cueing is when an experiment uses a written frame written that will "expose people to metaphoric language or imagery intended to cue a particular mapping" (Keefer & Landau 2016: 398). The explicit approach cues participants to use a metaphorical frame to understand a source domain through its target (Galinsky & Glucksberg 2000; for an overview see Landau 2017; Thibodeau et al. 2017). One popular type of study that uses explicit framing, pertinent to this chapter, is that which considers the role of metaphor in situations of prediction and problem-solving.

Studies using explicit metaphor framing in problem-solving scenarios cover a wide array of issues. As a demonstration of the influence of metaphor in prediction, a 2007 study had participants reason about an inanimate force (namely,

---

[5] The term *frame* as defined here is generally broad for the purposes for this article. For further disambiguation, see Fillmore and Baker (2009).
[6] I use the word "wanted" here quite loosely in that it doesn't necessarily require motivated use to frame an issue.
[7] Keefer and Landau refer to this divide as *conventionality*, a spectrum that describes the explicitness by which comparisons are made between source and target domains in a particular mapping. As formulated by Keefer and Landau, "The mappings in metaphor are conventional in everyday thought and language and, as such, can be activated automatically and with little effort. Drawing novel or unconventional mappings in analogy, on the other hand, is relatively more deliberate or difficult" (Keefer & Landau 2016: 396). For clarity, I reduce this term to the explicit/implicit cueing function.

the economy), framed to be understood as either an inanimate object, subject to outside forces, or as an agentive force. To put this more formally, participants were given an "object causality schema" and a contrasting "action schema" (Morris et al. 2007: 176). The working hypothesis was that "object causality schemas trace movements primarily to external forces, whereas action schemas trace movements to enduring internal properties. Hence action schemas create a bias to expect that observed trends will continue" (Morris et al. 2007: 176). The participants – investors, in this case – were asked whether they thought a trend in the stock market would continue, after having been given a description of the market along with some graphs. Participants were asked to read texts framed with metaphoric phrases describing an agentive market trend, like "jumped," "climbed," "wandered," while others read materials using non-agentive terms like "swept" or "plummeted." With these frames, the market can be seen to be either an active agent in itself or subject to outside forces. The investors then predicted what they thought the next day's market would look like. As hypothesised, "expectancies of trend continuance were higher in the agent-metaphor condition than elsewhere" (Morris et al. 2007: 178) even when the graphs were the same.

In a now-classic study of metaphor and problem solution, Thibodeau and Boroditsky (2011) conducted a study where crime was framed either as a beast or as a virus. The participants' solutions of an appropriate response to the crime varied drastically, even when given exactly the same crime statistics. As the study reports:

> Crime reducing suggestions differed systematically as a function of the metaphor used to frame the crime problem. Participants who read that crime was a virus were more likely to propose treating the crime problem by investigating the root causes of the issue and instituting social reforms than participants who read that crime was a beast. Participants who read that crime was a beast were more likely to propose fighting back against the crime problem by hiring police officers and building jails – to catch and cage the criminals – than participants who read that crime was a virus. (Thibodeau & Boroditsky 2011: 5)

This kind of problem solution and ascription of attitudes towards societal issues has been repeated in a number of different scenarios, in both discourse analysis and in experimental scenarios. These have included: understanding of radiation treatment by way of an analogy to war (Gick & Holyoak 1980; Holyoak & Koh 1987); choosing solutions to a war scenario by framing the choices in terms of past wars (highly supported war vs less supported war), among others (see Musolff 2004); the consequences for participant attitudes toward immigration when seen as an organism, an object, a natural catastrophe/war or as an animal (O'Brien 2003); determining whether counter-terrorism measures are more accepted when seen as war, as law enforcement, as containment of a social epidemic, or as a process

(Kruglanski et al. 2007); health, whether participants are more accepting of bad outcomes if a treatment of cancer is described as a journey or as a battle (Hauser & Schwarz 2015) or how metaphor affects understanding of randomization and chance in clinical trials (Krieger et al. 2010); or in trade, whether tariffs are preferred in situations where trade is described as a war or as a two-way street (Robins & Mayer 2000). Keefer and Landau propose a formalisation of these studies in the *metaphoric fit* hypothesis. Their hypothesis claims that "when the metaphoric framing of the solution 'fits' the metaphoric framing of the problem, people should prefer the solution, despite the fact that neither metaphoric framing is literally true" (Keefer & Landau 2014: 400). The clearest (and perhaps most colourful) example of these effects was a demonstration in the health domain, examining participants' evaluation of depression medications (Keefer et al. 2014). Participants were charged with evaluating two depression medications "Liftix" and "Illuminix." In one set, participants read descriptions of depression either framed vertically (where depression is described being physically *down* in various guises), or framed in terms of illumination (depression is being a *dark* period in one's life). Descriptions of each candidate drug was framed respectively, Liftix "(e.g. "has been shown to *lift* mood"; "patients everywhere have reported feeling *uplifted*" (Keefer et al. 2014: 14), and Illuminix (e.g. "has been shown to *brighten* mood"; "patients everywhere have reported a *brighter* outlook") (Keefer et al. 2014: 15). These were tested both against each other, and against neutrally framed descriptions. As hypothesised, participants – responding through a Likert scale measurement of the candidate solutions – preferred the metaphorically fit solutions.

Given the relevance for both national and international courts of the issues examined in the metaphoric framing literature, and their implication for legal reasoning, I argue that developing similar tests, along the framework laid out in section three, will go a long way towards replicating and – perhaps more importantly – falsifying some of the same evidence in a judicial setting, where other factors may be (and are often posited as) more determinative. Whether or not metaphoric fit is at work in legal reasoning has yet to be verified, particularly in an empirical setting. Prior work of the author has, however, indicated that it may indeed be present (Slosser 2019). The 2019 study aimed to establish whether the phenomenon of metaphoric fit may be at work in the case law of the European Court of Human Rights; in particular, in a chain of precedent in cases concerning a right to private life under Article 8. It concerned the framing of the concept of "the quality of law" in protection of privacy rights. It used paragraph-to-paragraph citation-chains to track how the phrase was understood, over a chain of precedent cases as cited by the Strasbourg court. Those paragraphs were then coded for metaphor-use, with respect to how terms such as quality, sufficiency, or adequacy were framed. In the studied cases, quality was understood through the STRUCTURE and VERTICALITY frames.

> When taken in the context of a solid-structure frame, any haphazard or irregular feature, any unbalanced measure, or non-independent functioning of duties causes the structure to fail or be subject to failure. In this way, the concrete image schema is used to attribute features of that schema to functions of the law. If it is a question of the quality of the law, we are reasoning with the embodied quality of strong structures. (Slosser 2019: 600)

This type of conventional metaphor might seem unsurprising to a legal linguist, but it is how these frames acted within the judgements themselves that is intriguing. Namely, how the judgment characterised competing arguments of the parties; and, which metaphors survived through a larger chain of cited precedent. This can be seen, for instance, in the cases from the study where the winning party's argument was described with frames congruent with the conclusion of the judgement, as opposed to a different, non-congruent frame of the non-winning party. Generally, the study found three types of congruencies between sets of metaphoric language that are consistent with the metaphoric fit hypothesis: first, as mentioned, between "[a] judgment and the competing arguments" as summarised by that judgment; second, between "[a] judgment and its cited precedent"; third, between "dissents to their own precedent choices and the rejection of the incongruent frames of the majority opinion" (Slosser 2019: 606–7). While compelling, the studies did not establish any causal relation between a framed precedent and its choice as a "good fit" among others. It also did not establish the influence of the written arguments of the parties on the selection of precedent,[8] it only established how they were represented by the judgement itself. Further, these were judgments written by multiple authors, so any cognitive constraint exerted through language is hard to establish. Shifting to an experimental design would help overcome such weaknesses. Much work is needed to create the kind of robust body of evidence, varied over a range of environments, participants, and legal subjects that has propelled metaphor study in the social psychology and psycholinguistics literature until now. The next section lays out the theoretical basis of the effect that metaphorical framing has on legal reasoning, and section three explains a practical approach to an experimental model for testing the metaphor fit hypothesis in the legal realm.

---

8 Experimental work of this nature is rare but not non-existent and most, if not all, is not concerned with the effects of metaphorical framing, see e.g. Salterio (1996); Spamann and Klöhn (2016).

2 What is at stake for legal reasoning?

If it is the case that metaphor can mediate our reasoning surrounding abstract concepts via other domains, and that understanding differs depending on how the concept is presented, it follows that law, as a principally linguistic discipline, would not escape the reach of Conceptual Metaphor Theory. Researchers need to answer a deceptively simple question: when does a change in language bring about a change in reasoning? Reasoning here should be understood in two senses. First, understanding the meaning of a term (i.e. a legal category) in interpretation. And second, reasoning about that meaning by use of analogy (e.g. using universal principles or reasoning from past cases). Conceptual Metaphor Theory can help us examine reasoning in both senses.

In the first sense, interpretation has traditionally relied on doctrinal categories. Building, defining, refining, excluding, and dividing these categories is how the law makes meaning, and therefore how its interlocuters reason about legal problems. From this point of view, legal interpretation is the act of legal meaning-making through defining a category's members, non-members, attributes, and contextual applications. Customarily, this is treated as a very dualistic endeavour, dividing things into that which belongs to the category and that which does not. Take for instance H.L.A. Hart's classic example of interpretation, when he asks about the meaning of the category *vehicle* for the purposes of a statute declaring "no vehicles allowed in the park" (Hart 1994 [1961]). To "find" meaning in a legal scenario like "is this a vehicle for the meaning of the statute" is to ask about a specific categorical distinction: is *x* a member of the legal category *y*? Meaning-making in this sense is the cognitive processes of allocating category membership, where that membership is allocated by a perceived similarity or difference between its members and potential members. Understood in this way, meaning in legal concepts is comparable to notions of structural mapping described in Conceptual Metaphor Theory. The working hypothesis for legal meaning in this formulation is: instance x is understood by its links (or lack of links) with the members of category *y* and those links are subject to how *x* and *y* are described, and therefore will be deemed as more or less fit. Unlike this view, the traditional approach to legal meaning applies categories as binary: is this thing a vehicle or is it not a vehicle? Although this is practical (certainly for something like the law that needs to arrive at a discrete decision), it is not realistic from a cognitive perspective. Categories are more than a binary membership appraisal. They are subject to blurred membership and radial relationships, with members sharing some but not all the same attributes. This fact about categories is not a recent discovery (Wittgenstein's notion of "family resemblance" (2010 [1953]), and yet legal categories are still applied in a binary way.

In the second sense, the law populates categories by treating like cases alike, often through the lens of abstract principles or previous decisions, thereby imbuing those categories with meaning. A category starts as a universal rule involving membership ("no vehicles") and is applied to a particular case (e.g. a garbage truck). The interpreter's job is to determine the fit of the particular to that universal, and deem it as a category member or not. Although even students in an introductory law seminar are capable of pointing out the nuance of applying the rule and reach for other principles or motives to understand the category "vehicles," and to shoehorn in what they intuitively think might be correct, the formalism of the law demands that the interpreter act as though the nuance of this decision is a distraction or non-starter. This is seen when the judge or jurist is motivated by the hyperbolic imperative that the law *must* be read a certain way, or that an interpretation is *clear*. Though even such judges or jurists might be under no illusions that something really must be read a certain way, the permanence of the written legal word has already done its work. The "clarity" involved in clear meaning is an attempt to make sharp that which is blurry, and to firm up the law with assertions of certainty and justification.[9]

These two senses are intrinsically linked. One may start from a set of background assumptions (be they philosophical, cultural, institutional, biological, or other), but once we are speaking of interpretation within a legal system, one cannot really make meaning in the first sense without the second sense, unless some internal coherence is sacrificed. However, this does not mean that the first sense is lost. The view from a legal-cognitive perspective is that these categories are first built from background knowledge and that the terms that make up and surround analogical reasoning are based on this background knowledge. More importantly, different descriptions of analogues will cue different background knowledge, thereby changing the dynamics of category-membership. This is what much of the work in legal-cognitive linguistics indicates. The view from the legal doctrinal account, on the other hand, relies on abstract notions derived from normative theories of what the law perhaps should do or, even more abstractly, what the law demands should be done. This indicates potential sympathy to the cognitive linguistic approach, premised on the insistence on the law's internal coherence. Decisions must somehow hang together.[10] What is missing from this

---

9 For a lengthier discussion of the need for this type of justification and the universalisation of principles, see Bankowski and MacLean (2006).
10 This is the justificatory effect of coherence in law. In this respect, see the discussions in, among others, Raz (1979); Kress (1984); Raz (1995); Marmour (2005).

formal legal account are the tools we can use to discern cognitive structures underlying those categories: the mechanisms of how they hang together, and particularly the mechanisms from the point of view of the thinkers themselves, as evidenced by the language they use or are affected by. Given the interconnectedness of these two forms of reasoning, we can look to the familiar tenet of reasoning from past cases to see both senses.

In reasoning from previous decisions, analogy plays a significant and complex role, where a "legal reasoner perceives a relevant similarity between the situation involved in some previous decision and the situation at issue in the instant case, and then uses the analogy between [the two] to argue that the instant case ought to be [or not to be] decided in the same way as the previous one" (Schauer & Spellman 2017: 251–2). Like conceptual metaphor, in using analogical reasoning, the reasoner *maps* the relation between the present and past cases. In the terms used in Conceptual Metaphor Theory, she *retrieves* the *source* (past) attributes and ports them to the *target* (current) case.[11] Given this functional similarity between metaphor and analogy, I follow Thibodeau et al. in presuming that "the terms metaphorical reasoning and analogical reasoning [can be] used interchangeably to describe how people use knowledge of one domain to talk and think about another" (2017: 852). The role of analogical reasoning in legal theory has been thoroughly debated.[12] While trying to determine likeness, a reasoner might list attributes shared between both domains by considering "the relevant differences and similarities among them" (Larson 2019: 672). Many legal theorists have attempted to construct logically-valid approaches to the use of legal analogy that help solve both what it means to share a likeness and when an analogy is normatively acceptable or warranted. What is missing from these debates is the evidence that empirical, experimental research could furnish about precisely what is mapped by reasoners, from a multitude of similarities, and ultimately what counts as a *good* analogy to the reasoner in practice, whether normatively warranted or not, and how those similarities are mitigated by factors that have been shown to affect this type of cognition – like the studies from Conceptual Metaphor Theory outlined in the final section.

One way to understand how this might be accomplished is to use a practical and simple scenario. Dan Hunter, who is conversant with the legal debate on analogy, similarly looks to cognitive science. He asks, why is it "that some

---

11 For a thorough treatment of analogy, see Gentner (2017).
12 Developing e.g. distinctions between derived attributes of scenarios, applicable rules, and standards that apply to both domains (MacCormick 2005; Alexander & Sherwin 2008), as well as classifying what it means for the source and target domains to share a likeness (Schauer 1987; Posner 1993; Brewer 1996; among various others).

analogies are better than others?" (Hunter 2004: 151). His approach, developed to help teach law students to use and understand analogical reasoning, deploys a hypothetical scenario developed from the oft-used *Adams v. New Jersey Steamboat Co* case.[13] The scenario involves a judge who must decide on an arguably undefined rule in a case that only has two precedents. Hunter asks us to imagine a man who is on an overnight ferry. During the course of his trip, his luggage is lost. Feeling that the ferry company should be held responsible, the man sues. The judge now has a decision to make about whether or not the company is liable. Complicating the issue, is the fact that there are only two precedents in this case for reference. Hunter lays out the precedents:

> The first precedent involved a hotel proprietor who was found liable for a guest's stolen luggage, since it was held that part of the contract of hospitality involved reasonably safe storage of the guest's belongings. The second precedent involved a railroad company, which was found not liable for the loss of the luggage of a passenger who travelled in a sleeper berth, on account of the contract being primarily for travel and not for lodging.
> (Hunter 2004: 153)

It falls to the judge to work out whether the ferry is more reasonably conceived of as a "floating hotel" or a "seagoing train" (Hunter 2001: 2000). To describe the constraints on the judge in formulating this mapping, Hunter outlines the approach of Holyoak and Thagard's "multiple-constraint model," of "purposive," "surface," and "structural" constraints (Holyoak & Thagard 1995: 5). The *purposive constraint* is what is curtailed by "the *purpose* for considering the analogy at all" (Hunter 2004: 166 original emphasis). As Duncan Kennedy writes "[l]egal reasoning is a kind of work with a purpose, and here (in adjudication) the purpose is to make the case come out the way my sense of justice tells me it ought to, in spite of what seems at first like the resistance or opposition of 'the law'" (Kennedy 1986: 526). This is the traditional notion of legal reasoning. The empirical testing of analogical reasoning in law in this way hits an obstacle, because it is difficult, if not impossible, to determine whether a legal reasoner's choice of precedent via analogical fit might be more curtailed by a purpose-driven constraint rather than by a surface or a structural constraint – particularly without knowing the baseline of either. Purpose may be motivated by a number of factors outside of analogical mapping of surface and structural elements. Surface constraints are those that display "*direct similarity* in the surface-level elements in the source and target domains" where it "is immediately possible to establish a one-to-one mapping between surface-level objects within the two domains" (Hunter 2004: 159, original emphasis). Whereas structural constraints are

---

[13] *Adams v. New Jersey Steamboat Co.*, 151 N.Y. 163, 45 N.E. 369 (N.Y. 1896)

those that "involve [. . .] the pressure to identify consistent structural parallels between the two domains" (Hunter 2004: 163). The difference involves the mapping not just of features between two domains, but also mapping the relations and systematicity between features.

In the ferry scenario, there are surface similarities mapped between both: a ferry and a hotel: "locked rooms [. . .] restaurants, bars; and so on" (Hunter 2004: 159); and a ferry and a train: "travelers moving between two places" (Hunter 2004: 159). The structural mapping comes from "the resemblance in the underlying systems of relations between the elements of the sources and the elements of the target" (Blanchette & Dunbar 2000: 108). For instance, a ferry's transport element entails a set of relations similar to the train's "transport" element, in that the relations: *passenger boards onto X, X moves from destination A to destination B, passenger buys ticket for X*, etc. where X can be filled by both the target (ferry) and source (train) domains. This similarity is indicative of an entire system of relations, whose degree of similarity and coherence as a system can strengthen or weaken the analogical fit of any given circumstances. All of these analogies would be quite familiar to most, if not all, legal scholars as explicit and purposeful mapping. What isn't clear is how this mapping may also rely on the constraints of language. Hunter points to this briefly in his discussion of context-effects of other precedents, but is silent on another potential factor – metaphor – in the analogies' perceived success. For this we can turn to the concept of conceptual metaphor for one unit of analysis, and in particular to the hypothesis of metaphoric fit as outlined in section one.

3 Formalising a testable model for Conceptual Metaphor Theory and legal reasoning

The proposed model has been developed as a way to test for the direct influence of these constraints in analogical reasoning in a legal setting. It uses metaphoric framing of precedents and the wording of a judgment to test consistency with the literature in Conceptual Metaphor Theory; in particular, the metaphoric fit hypothesis. First, one can isolate the effect of framing of precedents from the influence of arguments by removing arguments as features of the model.[14] Guiding the model are the congruency claims found in both Slosser (2019) and the literature outlined in section one. Those claims formally stated are:

---

14 Further studies would add this level of analysis. See section 4 below.

Claim one: in choosing a case for an analogous precedent, participants will prefer those that are framed congruently to the framed case, as compared to neutral and incongruently framed cases.

Claim two: these effects will be consistent among different sets of metaphorical frames.

The model will use the example from Hunter (2001) above, with the scenario of the train and ferry, to concretise the explanation. In this scenario, the participants would be presented with the test case *(T)* as a short paragraph describing the scenario of the man and his luggage. They will then be presented with each of the precedent cases, *(P)* – the precedent with the positive liability outcome, the hotel; and *(N)* – the precedent with the negative liability outcome, the train.

T = Ferry
P = Hotel (positive, meaning found liable)
N = Train (negative, meaning found not liable)

For example, the test case text *(T)* might read (slightly adapted from Hunter 2004):

> A man is suing for the loss of his luggage while he was aboard an overnight ferry. The luggage was stolen from the plaintiff's berth. The plaintiff is suing the ferry company, on the theory that it was responsible for his bags and therefore liable for their loss.
>
> As the judge you must decide whether or not the company is liable. Let us assume that there is no statutory pronouncement on the subject, and there are only two prior cases that might be relevant to the decision.

The participant would then be presented with the prior case texts of the hotel and train (*P* and *N*). Much like the depression texts and the candidate treatment described in the Keefer et al. (2014) study, the experiment would rephrase sentences to highlight conceptual metaphors belonging to different frames. Depending on the group of the participant (see below), the texts will be written with either or a neutral/literal frame *(O)* as above, or a metaphoric frame *(a, b, c, d, . . .)*. These frames are represented through subscript. For instance, the above neutrally framed test case would be represented as T_O. Similarly, one could use the frame of MOVEMENT *(a)* to cue structural similarities between the ferry and a train, which could use conceptual metaphors like BEING ON A FERRY IS BEING ON A JOURNEY, where elements of a journey and movement would be highlighted. Text T_a might read (emphasis added):

> A man is suing for the loss of his luggage while he was *travelling* aboard an overnight ferry *crossing* the North Sea. The luggage was stolen *during the course of his trip* from a *luggage* rack in the plaintiff's *compartment*. The plaintiff is suing [. . .]

Similarly, one could use a HOSPITALITY frame *(b)* to cue structural similarities between the ferry and a hotel which might make use of a metaphor of A FERRY IS A HOME, T_b would read:

> A man is suing for the loss of his luggage, while *staying in* an overnight ferry *in* the North Sea. The luggage was stolen from the overhead *storage* where *it was housed* in the plaintiff's *room*. The plaintiff is suing [. . .]

The prior cases would similarly be framed (P_a, P_b, N_a, N_b) or neutral (P_0, N_0) and presented to the participant. The experiment would focus on one frame group at a time *(a, b)*, *(c, d)*, etc. The same scenarios with different frame groups would be used in either the same study or in follow-up experiments. This would ensure that any effects found were due to congruency rather than the strength of one particular metaphoric frame or evoking some background knowledge that is more determinative to the outcome. The experiment results in 9 possible texts that the researcher would have to create:

$$T_a, T_b, T_0, P_a, P_b, P_0, N_a, N_b, N_0$$

These texts can then be constructed in all required arrangements, to test the effect of metaphoric framing on analogy making and precedent choice. For instance, the arrangement $T_a (N_a P_0)$ would represent the test case T, written with frame *a* (T_a), the negative precedent candidate N with frame *a* (N_a), and the positive precedent candidate P with a neutral frame 0 (P_0). In our ferry example, *a* represents the ferry in terms of movement, highlighting the transport element as a structural constraint (T_a). The metaphoric fit hypothesis would suggest, if similarly framed – as it is in this configuration with frame *a* – that the negative train precedent (N_a) would be expected to be a more salient "candidate solution," to which precedent applies. Thus, the participant would choose a congruently framed train where the company was not found liable (N_a) as a more analogous precedent when placed against a literally framed hotel who was found liable (P_0). This would be measured by first asking the participant to make a choice about how to rule, and then presenting the participant with a Likert scale to measure how the participant rated how well each prior case fit the test case.

The hypothesis would yield an expected outcome of a judgment (J) where the ferry company is found not liable $(_N)$. This could be represented by J_{Na}. The resulting arrangement would be: $T_a (N_a P_0) J_{Na}$. Following this logic, the remaining arrangements make up the text groups in which participants will be randomly placed to test the claims outlined above: *Isolated*, a congruently framed precedent is tested against a neutrally framed counterpart, testing claim one; *Direct*, both candidate precedents framed in direct competition against one another, testing a variant of claim one; and, *Base*, where both candidate precedents are neutrally framed and paired with a framed test-case; and, the control baseline, where all texts are neutrally framed.

Isolated – a congruently framed precedent is tested against a neutrally framed counterpart. This tests claim one.		**Direct** – both candidate precedents are framed in direct competition against one another. Tests a variant of claim one		**Base** – both candidate precedents are neutrally framed and paired with a framed test case; and, the control baseline where all texts are neutrally framed	
Text group	*Expected result*	*Text group*	*Expected result*	*Text group*	*Expected result*
$T_a (N_a P_o)$	J_{Na}	$T_a (N_a P_b)$	J_{Na}	$T_a (N_o P_o)$	baseline
$T_b (N_o P_b)$	J_{Pb}	$T_b (N_a P_b)$	J_{Pb}	$T_b (N_o P_o)$	baseline
				$T_o (N_o P_o)$	baseline

Aside from its primary function to measure deviation from a baseline, an added benefit of the base groups as arranged here is that it they might reveal the effects of preconceived notions of the participants. In the above example from Hunter, one such preconceived notion that may have affected the results is the perceived similarity between ferries, hotels and trains. The seven arrangements used here do not represent all the possible arrangements of the variables, but they represent those best suited to testing claim one.[15] These arrangements allow for the testing of all relevant permutations, while they help account for (without ruling out) prior beliefs about the candidate precedents (i.e. the purposive constraint), while simultaneously testing claim two. A third claim could also be tested here, by adding a section on the participants' reasoning on the case. This would test yet another initial finding from previous work: the finding that, in reasoning about the likeness between cases, participants will adopt in their explanations a framing congruent with the cited precedent in their explanations.

---

15 First, each of the first two sets can be flipped so that incongruent frames would be tested in isolation with a neutral frame. As this arrangement does nothing to test the metaphoric fit hypothesis, it does not need to be used. Second, a flipped direct set where the structural constraint frames would be switched. A flipped version would invert these framings: frame *a*, for the hotel, and frame *b*, for the train. Instead of testing by flipping the direct set, adding a second frame pair could accomplish testing this by changing the frames completely *(c, d)* and inverting the negative and positive judgment outcomes (i.e. the hotel company will now not be held liable, and the train company will be held liable).

4 Building a body of evidence and implications for future research

The intricacies of legal reasoning have been approached by a number of disciplines. The view offered here provides that, given the evidence from cognitive linguistics, there is every reason to think that metaphor, as understood in this model, influences legal reasoning, even while that influence may be mitigated by other factors. Such a view is important to understanding legal reasoning, particularly the type that involves analogical thinking, including reasoning from past cases. This discussion has proposed experimental means of securing evidence of reasoning rather than only evidence of patterns of language use in legal genres (typically corpus evidence). The claim that metaphor influences legal reasoning will remain untestable until there is a body of experimental evidence of the baseline effect of metaphor. Future research based on this model could look at phenomena supplementary to the main testable claims. Adjustments made to the model, along the parameters of legal area, participant demographics, testing conditions, and others, would both mark the prevalence of this linguistic phenomenon and begin to measure the degree of its impact among other factors. This concluding section will quickly look at a few of these possible adjustments and their potential implications for future research.

There are numerous legal issues that could immediately benefit from and be used to explore this experimental model. For instance, one could explore the context of causal factors in a tort claim for recovery of losses or damages, to determine, for example, whether there is a causal link between an agent and losses or if the loss is due to market factors. Rather than strictly trying to solve the puzzle of legal causation (see Swisher 2007; Haley 2000), there may be linguistic factors that strongly affect the reasoning of judges about causation. Recalling the Morris 2007 study about object vs agent causality, the researcher could use the above model to test conceptual metaphor's influence on attributing legal casualty in tort claims. This needn't be restricted to just traditional tort claims, but could be applied to legal causation and responsibility in a wider sense. Slightly altered again, future scenarios involving autonomous vehicles or algorithmic decision-making systems and responsibility could be explored. Here the researcher might set up the model to frame a medical decision, or an administrative decision between a human decision-maker and a machine, each being framed with agent vs object-oriented causality. It would cue participants to think of the decision-maker as a *cog in the machine* (so to speak) vs the decision-maker having agentive qualities. The assessment in this case might hinge on whether a system is seen as a tool used by a human or as an agent in itself. It could be that background knowledge about machines and humans is in itself too much for metaphoric framing to overcome, but early results from a current experiment similar to this scenario, conducted by the author, points to metaphor still having the expected effect. These, of course, are not the only two legal areas in which one could use the model, but are indicative of the kinds of adjustment one could make to the experimental method in order to expand its applicability.

A comparable adjustment would be to adjust the kind of solution-finding scenario in the model. For example, in many legal contexts, a court is saddled with a proportionality test that involves weighing whether a regulator or state's infringement on some right was deemed "necessary" by way of an "least restrictive means test." That is, was the manner in which the government attempted to solve some societal problem the one that would inflict the least harm on others? (see Sykes 2003). This is often a difficult and quite ambiguous assessment, and problem solution of this sort may also be subject to effects of conceptual metaphor. Like the solution attribution evidence presented in section one, the researcher could present the participant with framed solutions to decide which would be more appropriate to a given scenario as the balanced approach. This could first come with obvious frames of weight or balance as discussed earlier, but could also expand to more abstract framings such as the beast, or virus, or war metaphors from the Conceptual Metaphor Theory literature, depending on the scenario chosen. These types of adjustments would test the how metaphor might mediate a proportionality assessment. Going further, the experimenter could adjust the model to explore other established types of decision-making affected by metaphoric framing not covered explicitly in this chapter, such as the role of metaphor and memory, moral judgement, norm maintenance, and collective thought (for an overview, see Landau 2017) among others.

Adjustments could also be made to the testing environment and texts for other factors involved in decision making. For example, one could make the study more longitudinal and add further cases over time and test instances of the "path" of metaphoric framing through chains of precedent. It would ask questions such as: how long does a dominant frame hold up in a chain of precedent? How wide is its effect? The author's prior work hints at other phenomena from cognitive linguistic theory that can be seen in tracking concepts through a chain of cited cases. As a body of case law built up with multiple judges and decisions, if the metaphoric fit hypothesis is correct about reasoning from precedent, some frames would become more dominant than others, and there would be instances where a judge would need to choose between two dominant paths of case law that might both seem highly similar to the case at hand. Here the cognitive linguistic literature would point to the role of conceptual blending in merging differently framed precedents to create a blended frame that is then cited in the current case, instead of just reproducing one frame. Conceptual blending, in short, "aligns two partial structures (the inputs)" like the structure mapping of metaphor and analogy, "but in addition [. . .] projects selectively to form a third structure, the blend" (Fauconnier 2001: 256; see also Fauconnier & Turner 2008; Coulson 2001). When judging the similarity between test and prior cases the judge may import like attributes from both strands, creating a frame hybrid or blend that now becomes more dominant in the case law. Evidence of this was apparent in the wider study to which Slosser (2019) contributed. Going beyond the baseline of the model proposed here would aim to test these kinds of dynamics by building beyond the single case, two-precedent model.

Similarly, the aforementioned role of submitted pleadings by the parties might play a larger role than reasoning from previous cases. The model could be adjusted to test the effect of arguments over a baseline of previous case effects, framing the original arguments to see if the frames are reproduced by the judgment, or if the arguments are reframed to support the conclusion of the judge testing long-argued ideas of causal reasoning vs post-hoc rationalisation. Likewise, these adjustments could stretch to group dynamics by multiple judges, the effect of levels of expertise of the interpreter, differences in translation and multi-lingual environments, among others. The numerous possible ways of rearranging this simple framework to different scenarios will allow a solid body of evidence to be put up against other external factors that affect judicial decision-making. For instance, much of the literature dedicated to the behaviour of judges focuses on external factors, in the form of institutional mechanisms (Hall 2018) multivariate analysis (Skiple et al. 2016), or infamously when judges eat lunch (Danziger 2011). Other literature focuses on the motivations of judges in why they might decide the way they do on any number of legal issues, either as internal to their reasoning due to personal characteristics (see e.g. Braman 2009) or to ensure compliance with judgments (e.g. Dothan 2011). Another (massive) batch of literature ports the findings from behavioural economics – for example cognitive biases, heuristics, prospect theory, and the like – to explain judicial behaviour (see the research in the US stemming from Jolls et al. 1998; and the more-sparse EU law literature, Hacker 2015). This range of inquiry is breathtakingly diverse and justifiably influential.[16] Once the baseline is established, multi-factorial experiments can introduce many of these factors to test the claims presented in this chapter even more thoroughly.

---

[16] Although influential, the gap between the cognitive actor and the legal writer and reader in much of this literature remains. This gap is closing due to a general turn towards experimental methods, but without much import for legal philosophy or legal theory in the grander sense (Chilton & Tingley 2013; van Aaken 2020).

Adjustments such as these could also help tackle some of the issues debated within law and linguistics. Chief among these would be to test the methods developed in corpus linguistics to derive "ordinary meaning"[17]: using a larger corpus to distinguish the meaning of words. The main import of the corpus-based approach in deriving ordinary meaning is to use lexeme frequency (which could incorporate anything from a single word to a phrase) and its collocational attributes (what words surround a particular lexeme) to determine how a word has been used and might be properly understood.[18] The model could be adjusted, for instance, by collecting common collocations for a target term, and using different framings to test if the meaning derived from corpus-based methods hold up against metaphoric framing in an experimental setting. This might go a long way towards breaching the debates between semantics and pragmatics in an experimental, empirical fashion (see, generally, Pirker & Smolka 2019; 2020). There are numerous other concepts even within cognitive linguistics that could have enormous instrumental import for legal thought, such as: prototype theory (Rosch 1975a; Rosch 1975b; Rosch & Mervis 1975; Rosch & Lloyd 1978), conceptual blending (Fauconnier & Sweetser 1996; Fauconnier & Turner 2008), or idealised cognitive models (Cienki 2007) among others.

Regardless of the adjustments used, a framework like the one proposed in this chapter fills the need of legal linguistic and research into legal thought to balance its other empirical efforts with a falsifiable, repeatable method.

References

Aaken, Anne van. 2019. Experimental insights for international legal theory. *European Journal of International Law* 30 (4). 1237–1262.

Alexander, Larry & Emily Sherwin. 2008. *Demystifying legal reasoning*. Cambridge, UK: Cambridge University Press.

Bankowski, Zenon & James MacLean. 2006. Introduction. In Zenon Bankowski (ed.), *The universal and the particular in legal reasoning*, xi–xx. Farnham, UK: Ashgate Publishing, Ltd.

---

17 An effort at the foundation of legal interpretation, regardless of geography, see Slocum (2015); Zarbiyev (2015).

18 While it has its uses, the corpus approach is not without its criticisms. Mainly, "that ordinary meaning diverges from ordinary use" (Tobia 2020: 735). As laid out in this chapter, there is more to the puzzle of meaning than ordinary meaning. Work in other strands of legal linguistics such as what has been described the "pragmatic turn" in law focuses primarily on the ways in which context can inform how hearers understand implied meanings (see Giltrow & Stein 2019).

Bianchi, Andrea. 2016. *International law theories: An inquiry into different ways of thinking*. Oxford: Oxford University Press.
Blanchette, Isabelle & Kevin Dunbar. 2000. How analogies are generated: The roles of structural and superficial similarity. *Memory & Cognition* 28 (1). 108–124.
Braman, Eileen. 2009. *Law, politics, and perception: How policy preferences influence legal reasoning*. Charlottesville: University of Virginia Press.
Brewer, Scott. 1996. Exemplary reasoning: Semantics, pragmatics, and the rational force of legal argument by analogy. *Harvard Law Review* 109 (5). 923–1028.
Chilton, Adam & Dustin Tingley. 2013. Why the study of international law needs experiments. *Columbia Journal of Transnational Law* 52 (1). 173–237.
Cienki, Alan. 2010. Frames, idealized cognitive models, and domains. In Dirk Geeraerts & Hubert Cuyckens (eds.), *The Oxford Handbook of Cognitive Linguistics*, 170–187. Oxford: Oxford University Press.
Cotterill, Janet. 2010. How to use corpus linguistics in forensic linguistics. In Anne O'Keeffe & Michael McCarthy (eds.), *The Routledge handbook of corpus linguistics*, 606–618. London: Routledge.
Coulson, Seana. 2001. *Semantic leaps: Frame-shifting and conceptual blending in meaning construction*. Cambridge, UK: Cambridge University Press.
Croft, William & D. Alan Cruse. 2004. *Cognitive linguistics*. Cambridge University Press.
Danziger, Shai, Jonathan Levav & Liora Avnaim-Pesso. 2011. Extraneous factors in judicial decisions. *Proceedings of the National Academy of Sciences*. National Academy of Sciences 108 (17). 6889–6892.
DiMaggio, Paul. 1997. Culture and cognition. *Annual Review of Sociology* 23 (1). 263–287.
Dothan, Shai. 2011. Judicial tactics in the European court of human rights. *Chicago Journal of International Law* 12 (1). 115–142.
Dunoff, Jeffrey L. & Mark A. Pollack. 2017. Experimenting with international law. *European Journal of International Law* 28 (4). 1317–1340.
Evans, Vyvyan. 2007. *A glossary of cognitive linguistics*. Edinburgh: Edinburgh University Press.
Evans, Vyvyan & Melanie C. Green. 2006. *Cognitive linguistics: An introduction*. Edinburgh: Edinburgh University Press.
Fauconnier, Gilles. 2001. Conceptual blending and analogy. In Dedre Gentner, Keith James Holyoak & Boicho N. Kokinov (eds.), *The analogical mind: Perspectives from cognitive science*, 255–285. Boston: MIT Press.
Fauconnier, Gilles. 2006. Pragmatics and cognitive linguistics. In Laurence R. Horn & Gregory Ward (eds.), *The Handbook of Pragmatics*, 657–674. Oxford: Blackwell.
Fauconnier, Gilles & Eve Sweetser. 1996. *Spaces, worlds, and grammar*. Chicago: University of Chicago Press.
Fauconnier, Gilles & Mark Turner. 2008. *The way we think: Conceptual blending and the mind's hidden complexities*. New York: Basic Books.
Fillmore, Charles J. & Collin Baker. 2009. A frames approach to semantic analysis. In Bernd Heine & Heiko Narrog (eds.), *The Oxford handbook of linguistic analysis*, 792–816. Oxford: Oxford University Press.
Galinsky, Adam D. & Sam Glucksberg. 2000. Inhibition of the literal: Metaphors and idioms as judgmental primes. *Social Cognition* 18 (1). 35–54.
Gentner, Dedre. 2017. Analogy. In William Bechtel & George Graham (eds.), *A companion to cognitive science*, 107–113. Oxford: Blackwell.

Gick, Mary L. & Keith J. Holyoak. 1980. Analogical problem solving. *Cognitive psychology* 12 (3). 306–355.
Giltrow, Janet & Dieter Stein (eds.). 2019. *The Pragmatic turn in law: Inference and interpretation in legal discourse.* Boston & Berlin: De Gruyter.
Hacker, Philipp. 2015. The behavioral divide: A critique of the differential implementation of behavioral law and economics in the US and the EU. *European Review of Contract Law* 11 (4). 299–345.
Haley, Peter C. 2000. Paradigms of proximate cause. *Tort & Insurance Law Journal* 36 (1). 147–165.
Hall, Melinda Gann. 2018. Decision making in state Supreme Courts. In Robert M. Howard & Kirk A. Randazzo (eds.), *Routledge Handbook of Judicial Behavior*, 301–320. London: Routledge.
Hamann, Hanjo & Friedemann Vogel. 2017. Evidence-based jurisprudence meets legal linguistics: Unlikely blends made in Germany. *BYU Law Review* 2017 (6). 1473–1501.
Hart, H.L.A. 1994. *The concept of law*. Oxford: Clarendon Press.
Hauser, David J. & Norbert Schwarz. 2015. The war on prevention: Bellicose cancer metaphors hurt (some) prevention intentions. *Personality and Social Psychology Bulletin* 41 (1). 66–77.
Holyoak, Keith J. & Kyunghee Koh. 1987. Surface and structural similarity in analogical transfer. *Memory & cognition* 15 (4). 332–340.
Holyoak, Keith J. & Paul Thagard. 1995. *Mental leaps: Analogy in creative thought*. Boston: MIT Press.
Hunter, Dan. 2001. Reason is too large: Analogy and precedent in law. *Emory Law Journal* 50 (4). 1197–1264.
Hunter, Dan. 2004. Teaching and using analogy in law. *Journal of the Association of Legal Writing Directors* 2 (1). 151–168.
Ignatow, Gabriel. 2007. Theories of embodied knowledge: New directions for cultural and cognitive sociology? *Journal for the Theory of Social Behaviour* 37 (2). 115–135.
Jolls, Christine, Cass R. Sunstein & Richard Thaler. 1998. A behavioral approach to law and economics. *Stanford Law Review* 50 (5). 1471–1550.
Jostmann, Nils B., Daniël Lakens & Thomas W. Schubert. 2009. Weight as an embodiment of importance. *Psychological science* 20 (9). 1169–1174.
Keefer, Lucas A. & Mark J. Landau. 2016. Metaphor and analogy in everyday problem solving: Metaphor and analogy in everyday problem solving. *Wiley Interdisciplinary Reviews: Cognitive Science* 7 (6). 394–405.
Keefer, Lucas, Mark Landau, Daniel Sullivan & Zachary Rothschild. 2014. Embodied metaphor and abstract problem solving: Testing a metaphoric fit hypothesis in the health domain. *Journal of Experimental Psychology* 55. 12–20.
Kennedy, Duncan. 1986. Freedom and constraint in adjudication: A critical phenomenology. *Journal of Legal Education* 36 (4). 518–562.
Komarek, Jan. 2013. Reasoning with previous decisions: Beyond the doctrine of precedent. *American Journal of Comparative Law* 61 (1). 149–172.
Kövecses, Zoltán. 2017. Levels of metaphor. *Cognitive Linguistics* 28 (2). 321–347.
Kress, Kenneth J. 1984. Legal reasoning and coherence theories: Dworkin's rights thesis, retroactivity, and the linear order of decisions. *California Law Review* 72 (3). 369–402.
Lakoff, George. 2008. *Women, fire, and dangerous things*. Chicago: University of Chicago Press.
Lakoff, George & Mark Johnson. 1980. *Metaphors we live by*. Chicago: University of Chicago Press.
Larson, Brian. 2019. Law's enterprise: Argumentation schemes & legal analogy. *University of Cincinnati Law Review* 87 (3). 663–721.

MacCormick, Neil. 2005. *Rhetoric and the rule of law: A theory of legal reasoning*. Oxford: Oxford University Press.
MacCormick, Neil & Robert Summers (eds.). 1997. *Interpreting precedents: A comparative study*. Farnham, UK: Ashgate Publishing.
Marmor, Andrei. 2005. *Interpretation and legal theory*, 2nd edition. Oxford & Portland, OR: Hart Publishing.
McAuliffe, Karen. 2020. Creating multilingual law. In Malcolm Coulthard, Alison May & Rui Sousa-Silva (eds.), *The Routledge Handbook of Forensic Linguistics*, 64–78. London: Routledge.
Morris, Michael W., Oliver J. Sheldon, Daniel R. Ames & Maia J Young. 2007. Metaphors and the market: Consequences and preconditions of agent and object metaphors in stock market commentary. *Organizational Behavior and Human Decision Processes* 102 (2). 174–192.
Musolff, Andreas. 2004. *Metaphor and political discourse: Analogical reasoning in debates about Europe*. London: Palgrave Macmillan.
Pirker, Benedikt & Jennifer Smolka. 2019. The future of international law is cognitive: International Law, cognitive sociology and cognitive pragmatics. *German Law Journal* 20 (4). 430–448.
Pirker, Benedikt & Jennifer Smolka. 2020. International law and linguistics: pieces of an interdisciplinary puzzle. *Journal of International Dispute Settlement* 11 (4). 501–521.
Posner, Richard A. 1993. *The problems of jurisprudence*. Boston: Harvard University Press.
Raz, Joseph. 1979. *Authority of law: Essays on law and morality*. Oxford: Oxford University Press.
Raz, Joseph. 1995. *Ethics in the public domain: Essays in the morality of law and politics. ethics in the public domain*. Oxford: Oxford University Press.
Raz, Joseph. 2009. *The authority of law*. Oxford: Oxford University Press.
Robins, Shani & Richard E. Mayer. 2000. The metaphor framing effect: Metaphorical reasoning about text-based dilemmas. *Discourse Processes* 30 (1). 57–86.
Rosch, Eleanor. & Barbara Bloom Lloyd (eds.). 1978. *Cognition and categorization*. Hillsdale, NJ: Lawrence Erlbaum Associates.
Rosch, Eleanor. 1975a. Cognitive reference points. *Cognitive psychology* 7 (4). 532–547.
Rosch, Eleanor. 1975b. Cognitive representations of semantic categories. *Journal of Experimental Psychology: General* 104 (3). 192.
Rosch, Eleanor & Carolyn B. Mervis. 1975. Family resemblances: Studies in the internal structure of categories. *Cognitive Psychology* 7 (4). 573–605.
Salterio, Steven. 1996. The effects of precedents and client position on auditors' financial accounting policy judgment. *Accounting, Organizations and Society* 21 (5). 467–486.
Schauer, Frederick. 1987. Precedent. *Stanford Law Review* 39 (3). 571–605.
Schauer, Frederick & Barbara A. Spellman. 2017. Analogy, expertise, and experience symposium: Developing best practices for legal analysis. *University of Chicago Law Review* 84 (1). 249–268.
Skiple, Jon Kåre, Gunnar Grendstad, William R. Shaffer & Eric N. Waltenburg. 2016. Supreme court justices' economic behaviour: A multilevel model analysis. *Scandinavian Political Studies* 39 (1). 73–94.
Slocum, Brian G. 2015. *Ordinary meaning: A theory of the most fundamental principle of legal interpretation*. Chicago: University of Chicago Press.
Slosser, Jacob Livingston. 2019. Components of legal concepts: Quality of law, evaluative judgement, and metaphorical framing of Article 8 ECHR. *European Law Journal* 25 (6). 593–607.

Slosser, Jacob Livingston & Madsen Madsen Mikael Rask. Forthcoming. Institutionally Embodied Law: Cognitive Linguistics and the Making of International Law. In Moshe Hirsch & Andrea Bianchi (eds.), *International law's invisible frames: Social cognition and knowledge production in international legal processes*. Cambridge, UK: Cambridge University Press.

Solan, Lawrence M. 2020. Corpus linguistics as a method of legal interpretation: Some progress, some questions. *International Journal for the Semiotics of Law* (2). 283–298.

Spamann, Holger & Lars Klöhn. 2016. Justice is less blind, and less legalistic, than we thought: Evidence from an experiment with real judges. *The Journal of Legal Studies* 45 (2). 255–280.

Swisher, Peter Nash. 2007. Causation requirements in tort and insurance law practice: Demystifying some legal causation riddles. *Tort Trial & Insurance Practice Law Journal* 43 (1). 1–34.

Sykes, Alan O. 2003. The least restrictive means. *The University of Chicago Law Review* 70 (1). 403–419.

Talmy, Leonard. 2000. *Toward a cognitive semantics: Concept structuring systems*. Boston: MIT press.

Thibodeau, Paul H. & Lera Boroditsky. 2011. Metaphors we think with: The role of metaphor in reasoning. *PLoS One* 6 (2). 1–11.

Thibodeau, Paul H., Rose K. Hendricks & Lera Boroditsky. 2017. How linguistic metaphor scaffolds reasoning. *Trends in Cognitive Sciences* 21 (11). 852–863.

Tiersma, Peter M. 1999. *Legal language*. Chicago: University of Chicago Press.

Solan, Lawrence M. & Peter M. Tiersma. 2012. Introduction. In Peter M. Tiersma & Lawrence M. Solan, *The Oxford handbook of language and law*, 1–9. Oxford: Oxford University Press.

Tobia, Kevin. 2020. Testing ordinary meaning. *Harvard Law Review* 134 (2). 726–806.

Turner, Mark. 1994. *Reading minds: The study of English in the age of cognitive science*. Princeton, MA: Princeton University Press.

Van Aaken, Anne & Tomer Broude. 2019. The psychology of international law: An introduction. *European Journal of International Law* 30 (4). 1225–1236.

Vogel, Friedemann, Hanjo Hamann & Isabelle Gauer. 2018. Computer-assisted legal linguistics: Corpus analysis as a new tool for legal studies. *Law & Social Inquiry* 43 (4). 1340–1363.

Winter, Steven L. 2001. *A clearing in the forest: Law, life, and mind*. Chicago: University of Chicago Press.

Wittgenstein, Ludwig. 2010 [1953]. *Philosophical investigations*. G.E.M Anscombe, P.M.S. Hacker & Joachim Schulte (trans.). Oxford: Wiley-Blackwell.

Zarbiyev, Fuad. 2015. A genealogy of textualism in treaty interpretation. In Andrea Bianchi, Daniel Peat & Matthew Windsor (eds.), *Interpretation in international law*, 251–267. Oxford: Oxford University Press.

Jennifer Smolka and Benedikt Pirker
Pragmatics and the interpretation of international law: Two Relevance Theory-based approaches

Introduction

In recent years, researchers in various areas of law have begun to develop an interest in pragmatics. The present chapter focuses on international law and the interpretation of international treaties. We do not suggest that our findings apply exclusively to international law. We simply restrict ourselves in the following discussion for the sake of clarity, as there are differences in the terminology used in other fields of law – such as legal theory or specific domestic laws (common law vs. civil law systems, for example) – and between the different canons, rules and principles of interpretation that apply in various fields of law. There are certain very specific lessons that can be taken from pragmatics, in particular from Relevance Theory, as a cognitive-pragmatic approach. For this purpose, we first present certain features of interpretation in international law. We then turn to a brief presentation of Relevance Theory, and add the first useful application that can – and in our view should – be taken up by international lawyers, namely: typologies of inferred meaning. We mainly rely here on Ariel's (2016) work. Second, based on comments received at the International Language

---

Acknowledgements: Jennifer Smolka is a PhD Candidate at the Chair for English Linguistics, University of Fribourg. Benedikt Pirker is a Senior Lecturer at the Chair of European, International and Public Law, University of Fribourg. The authors thank Ziv Bohrer, Lando Kirchmair, Laurence Horn, Dieter Stein and Izabela Skoczeń, the participants in the 2019 International Language and Law Association Conference at UCLA and in the 2019 16th International Pragmatics Conference at Hong Kong Polytechnic University, for valuable comments on earlier versions of this article. Comments are welcome at benedikt.pirker@unifr.ch and jennifer.smolka@unifr.ch. Most of Section 3 of the present chapter is the combined product of both authors and based strongly on another forthcoming legal publication (Pirker and Smolka, *A linguistic-pragmatic approach to treaty silence and interpretation*, paper currently under review). Jennifer Smolka is the sole author of sections 3.1 and 4 of the present chapter.

Jennifer Smolka, University of Fribourg
Benedikt Pirker, University of Fribourg

https://doi.org/10.1515/9783110720969-007

and Law Association Conference of 2019, we examine the notion of explicature in legal interpretation, from a neo-Gricean and a post-Gricean – that is, relevance-theoretic – perspective. In the final section, we discuss our conclusions and point out possible directions for future research.

1 Interpretation in international law and linguistics

Interpretation is omnipresent in law. It is regulated in law by special norms, referred to – depending on the field of law – as maxims, principles, rules or canons. In international law,[1] for example, the cardinal rule of interpretation according to Article 31 (1) of the Vienna Convention on the Law of Treaties,[2] requires that "[a] treaty shall be interpreted in good faith in accordance with the ordinary meaning to be given to the terms of the treaty in their context and in the light of its object and purpose." There are numerous legal debates over this cardinal rule. International lawyers argue over how this rule relates to other rules of that Convention (Regan 2017), and to what extent canons and principles of interpretation continue to exist beyond the system of the Convention (Klingler et al. 2019), although the latter is recognised as customary international law binding on all interpreting agents[3] in international law.[4] For the present purposes, it is most important to note that these are binding rules that interpreting agents need to follow. For example, an interpreting agent must take some documents or treaties into account as context, but not others (Dörr 2018: 590). The question arises as to what extent this system of binding rules relies on everyday language comprehension. If it does, we can rely on linguistic findings and transplant them to the context of the interpretation of international treaties. In the following subsections, we first identify the semantic and pragmatic principles we want to apply. Second, we examine the legal

---

[1] We focus presently on the interpretation of treaties and leave aside the topic of the interpretation of customary international law (Merkouris 2017; Chasapis Tassinis 2020).
[2] Vienna Convention on the Law of Treaties, 23 May 1969, *United Nations Treaty Series*, Vol. 1155, p. 331.
[3] We will use the term *interpreting agents* to encompass all those called to interpret international law, such as international courts and tribunals, international organisations, states, as well as domestic courts and others (see Aust and Nolte 2016).
[4] As recognised by international courts, e.g. the International Court of Justice in *Avena and Other Mexican Nationals (Mexico v. United States)*, ICJ Reports 2004, p. 12; or the World Trade Organisation Appellate Body in *United States: Standards for Reformulated and Conventional Gasoline*, AB-1996-1, WT/DS2/AB/R, 29 April 1996, p. 17.

context of treaty interpretation to ensure that these principles "fit" that context. On this basis, we can then turn to two practical applications of pragmatics to legal interpretation, namely: pragmatic typologies of *inferred meaning*, and neo- and post-Gricean perspectives on *explicit meaning*.

1.1 Approaches to language meaning in linguistics: Semantics, pragmatics and Relevance Theory

To date, scholarship in international law has rarely relied in-depth on linguistics. If authors refer to anything linguistic at all when discussing interpretation, they typically cite Wittgenstein's idea of a language game and of language meaning as based on use (e.g. Klabbers 2017: 56). Only recently have a few scholars started to examine the semantics-pragmatics distinction, to see if findings in pragmatics can be made fruitful in the context of international law (Linderfalk 2007, 2013; Smolka & Pirker 2016; Pirker & Smolka 2017).

The two directions of research in semantics and pragmatics can be differentiated as being essentially about two different models for how communication works. Put very simply, semantics relies on a code model, which suggests that communication is encoded directly or indirectly in language. Pragmatics relies instead on an inferential model, according to which a communicator provides evidence of her intention in order to convey a meaning, and the audience infers this meaning on the basis of the evidence provided, on the contextual information and on their own prior knowledge (Wilson & Sperber 2006: 607; Zufferey & Moeschler 2012: 88). Whereas linguists do not agree to what extent communication can be explained by semantics or pragmatics (Horn 2006; Börjesson 2014), it is a fairly representative position to accept that both aforementioned models operate simultaneously and do not exclude one another. Verbal comprehension thus consists of the decoding of linguistic information as one input, but it does so in an inference process yielding an interpretation of a speaker's meaning (Moeschler 2009: 452). Take the example of the utterance "Can you pass me the salt?" Only part of the meaning can be arrived at by decoding. Without the situational context and a certain prior knowledge, it remains unclear for an addressee whether the interrogative sentence ought to be interpreted as a request or as a question, namely whether the addressee is physically able to pass the salt to the speaker.

A second important distinction – this time within pragmatics – also needs to be recalled, as it will play a major role in the following two applications of pragmatics to legal interpretation. This distinction opposes conventionalist to intentionalist approaches to language meaning. Its origins lie in Speech Act Theory (cf. Reboul & Moeschler 1998: 30).[5] The study of speech acts has shown that there is a significant gap between meaning as it is encoded in a linguistic form and the interpretation a speaker intends to convey to an addressee. While speech acts may go wrong – and in fact sometimes do go wrong – addressees are typically able to fill the gap in a seemingly effortless way. Therefore, scholars like Searle (1969) and Austin (1962) have suggested that there have to be systematic principles governing linguistic interaction. Grice (1975) developed the idea of a *cooperative principle* in this context. While they were interested in speakers' intentions in communication, "conventionalist" approaches in pragmatics rely on such principles – or "conventions" – as the basis of communication.[6] Current influential approaches that take up Gricean conventionalist concepts are often labelled as "neo-Gricean" (specifically, Hornian) approaches (Carston 2005: 307). While Grice and Horn are interested in speaker intentions, they do not focus on the actual processes of interpretation or the mental representations that a hearer or addressee forms of a speaker's or communicator's intention (Carston 2013: 9). Their goal is not to provide an account of "the processes of on-line utterance comprehension"; instead, they offer principles or maxims to account for utterance comprehension (Carston 2005: 305).[7]

"Post-Gricean" accounts, which are dominated by Relevance Theory (Sbisà 2006: 2223), argue that reliance on maxims or conventions is unconvincing. Namely, for communication to succeed under such an account, a speaker and an addressee would have to know that only shared assumptions are used in a communication process; the addressee would have to know that the speaker holds such an assumption, which again the speaker would have to know and so forth *ad infinitum*. Such mutual knowledge is thus impossible to achieve in the practice of verbal communication. Instead of this mutual knowledge requirement, Relevance Theory proposes that it suffices that discourse participants *assume* mutual assumptions (Sperber & Wilson 1995: 17–21). This notion is called *mutual manifestness* (Sperber & Wilson 1995: 41). For a fact to be *manifest* to an individual, it must be perceptible or inferable at a given moment. Since manifestness depends on the cognitive abilities and the physical and cognitive environments of an individual at a given moment (Sperber & Wilson, 1995: 39), an individual may attribute a similar manifestness to their interlocutor. That is, an individual may assume

---

[5] Speech Act Theory has also sometimes captured the interest of (international) lawyers (Klabbers 2017: 31; Fyfe 2017: 523).
[6] Grice's seminal contribution in this regard has recently also been taken up in law and legal interpretation in general (Macagno et al. 2018).
[7] See section 4 for more detail on Horn's approach.

there is a good chance that the *cognitive environments* – which are the sets of assumptions that are manifest to individuals (Moeschler 2009: 456) – of the two interlocutors overlap. To say that two people share a cognitive environment does, however, not imply "that they make the same assumptions: merely that they are capable of doing so" (Sperber & Wilson, 1995: 41), which also explains why communication may fail.

In line with a general turn towards cognitive approaches in the sciences, scholars have started to examine the mind, mental attitudes and intentionality (Sbisà & Turner 2013: 3), thereby pursuing a cognitively-grounded "intentionalist" approach, in the sense that the capacity to attribute mental states, such as intentions, is a capacity of the human mind (cf. Moeschler 2010: 222–223). Relevance Theory is one theory that has emerged from this intentionalist approach: it combines the study of utterance interpretation in context with elements of cognitive theory (Moeschler & Auchlin 2009: 178). Following the footsteps of Relevance Theory, scholars have developed typologies for pragmatic interpretations in order to systematise such interpretations. We will explain these typologies in more detail, and rely on them in the following section.

According to Relevance Theory, communication works on the basis of human cognition being constrained by a cognitive principle of *relevance*, in the sense of maximising relevance (Sperber & Wilson 1995: 261). In a situation like the one in our previous example of the utterance "Can you pass me the salt?", an addressee would therefore test interpretive hypotheses based on situational context and their own prior knowledge, to arrive at the most relevant interpretation (Moeschler 2009: 456; Sperber & Wilson 1995: 41, 44) – probably that of a request. The comprehension procedure follows a cost-benefit-logic, with processing effort as costs and cognitive effects – e.g. a change to the set of assumptions forming one's cognitive environment – as benefits (Wilson 2003: 282; Moeschler 2009: 456). Information is relevant if it has at least one *positive cognitive effect* in a given context. It is relevant if it adds, modifies, or deletes information (Moeschler 2009: 454). A positive cognitive effect is "a worthwhile difference to the individual's representation of the world" (Wilson & Sperber 2006: 620). In our example, the change in the addressee's assumptions consists in the addition of assumptions, for example, regarding the assumed wishes of the speaker (who would like to have salt) and their own need or capacity for action (to react to the request of the speaker and to provide salt).

After this brief introduction, we now turn to the normative framework of interpretation in international law. Before we can explain how the above-mentioned different models may be of help in sharpening our understanding of processes of legal interpretation, it must first be ascertained whether we are licenced to transplant these linguistic ideas into the legal context, that is, whether the relevant norms on interpretation allow it.

1.2 Language meaning, linguistics and the normative framework of interpretation in international law

The core norm of the international legal normative framework of treaty interpretation consists of Article 31 (1) of the Vienna Convention on the Law of Treaties.[8] This rule of interpretation prescribes that "[a] treaty shall be interpreted in good faith in accordance with the ordinary meaning to be given to the terms of the treaty in their context and in the light of its object and purpose." For our present purposes, there are two questions we need to examine. First, does this normative framework rely on everyday, non-specialised language comprehension (e.g. on the street or at the café) or does it rely on another specialised approach to communication? Only in the case of the former can we rely on semantics and pragmatics and transplant our linguistic findings into the context of treaty interpretation. Second, how does the normative framework deal with the intentions of the speaker? Only if the approach to speaker intentions is compatible with a semantic-pragmatic approach to language meaning, as set out before, can we rely on pragmatic theories in the following sections to provide explanations of linguistic phenomena in the interpretation of international law.

First, "ordinary meaning" plays a crucial role in treaty interpretation.[9] Most international lawyers would agree that as a first step in the process of interpretation, an interpreting agent, for example an international court, has to look at the text. As a representative view, one legal scholar maintains that ordinary meaning is to be established not as "any layman's understanding," but rather as what a person "reasonably informed on the subject matter of the treaty" would understand under a treaty's terms (Dörr 2018: 581). This description may be somewhat unsatisfactory, as it does not tell us any more exactly whose ordinary meaning is at issue (Slocum & Wong, forthcoming) – that of native speakers, or of a particular social group (e.g. English-speaking international lawyers), or of native speakers from a particular region (e.g. the United Kingdom and the United States, but not Australia)? Leaving this imprecision aside, we can nonetheless take as our starting point that ordinary meaning is to be established

---

[8] United Nations, Vienna Convention on the Law of Treaties, 23 May 1969, *United Nations Treaty Series*, Vol. 1155, p. 331.
[9] See in other legal contexts Slocum (2015), Bix (2012: 155).

through an ordinary process of a person understanding a written text, as long as there is a certain level of knowledge about the relevant treaty's subject matter. Several additional arguments can be developed that strengthen the case for the central role of this ordinary meaning rule. Notably, Article 32 of the Vienna Convention makes it possible to have recourse to a separate group of so-called "supplementary" means of interpretation to determine the meaning of a treaty provision. This would mean leaving aside ordinary meaning as prescribed in Article 31 of the Convention. However, Article 32 can only be used under the rare condition that the interpretation under Article 31 (i.e. relying on ordinary meaning) leaves the meaning ambiguous or obscure, or leads to a manifestly absurd or unreasonable result. Ordinary meaning thus remains crucial under all but the most exceptional circumstances. An interpretive agent must therefore always be able to explain and legally justify their interpretation in terms of the ordinary meaning of the relevant treaty provision.

Second, how does international law regulate the relevance of a speaker's intention? In our case of an international treaty, such an intention would correspond to that of the (state) parties that concluded the treaty. The normative framework on treaty interpretation foresees the establishment of the common intention of the parties as its overall objective (McNair 1961: 365; Bjorge 2014). The other elements of Article 31 (1) of the Vienna Convention also point in the same direction. Ordinary meaning is to be understood as the intention of the parties having found expression in a treaty text (Dörr 2018: 579–580). The International Court of Justice has emphasised that interpretation must be based "above all upon the text of the treaty" as an expression of the intention of the treaty parties;[10] and the International Law Commission similarly argues that the text is presumed to be the authentic expression of the intention of the parties.[11] As provided for in Article 31 of the Vienna Convention, context similarly involves the establishment of the common intention of the parties. As context, interpreting agents are permitted to have recourse to relevant subsequent practice of the parties to interpret the treaty. The reason for relying on such subsequent practice is that it reflects the parties' common intention (Sorel & Boré Eveno 2011: 826), which may have changed over time and may have resulted in a new, current consensus (Dörr 2018: 561). The object and purpose of a treaty is another means of interpretation in the Vienna Conven-

---

10 *Territorial Dispute (Libya v. Chad)*, Judgment of 3 February 1994, [1994] ICJ Reports 6, at 21, para 41.
11 Draft Articles on the Law of Treaties with Commentaries, 1966 Yearbook of the International Law Commission, Vol. 20 II, 220 para 11.

tion. Again, the intention of the parties plays a crucial role. In searching for this object and purpose, interpreters have to inquire about the intention of the parties; at the same time, they are bound by the limits of ordinary meaning of the treaty text in the sense that they must not introduce "independent sources of meaning" which contradict the text.[12] In this context, the principle of good faith (Linderfalk 2018: 29–30) also requires interpreting agents to focus on parties' intentions and prohibits the introduction of "through the back door" objectives that the parties to the treaty did not intend to be included in the terms of the treaty (Dörr 2018: 586). All in all, interpreting agents are thus to assume that there is an intention of the parties to a treaty. The "normal" process to establish this intention is to look at the text; only in exceptional cases, such as a special meaning intended by the parties under Article 31 (4) of the Vienna Convention (Sorel and Boré Eveno 2011: 829), or the consultation of the parties' subsequent practice, can an interpreter deviate from what is expressed in the text of a treaty.

Consequently, this overview shows that the normative framework of international law leaves us ample room to transplant concepts from semantics and pragmatics, as it relies both on ordinary meaning and on the intentions of the parties. This fits with the focus of semantics and pragmatics with regard to the extent to which meaning is encoded in language, as well as the focus on meaning as established through language use based on the intention of a speaker as inferred by an addressee. Our analysis does not attempt to change or replace the existing legal framework, nor are we in a position to do so. Rather, the goal of the following reflections is to propose a refinement for (international) legal thinking when dealing with certain pragmatic phenomena during processes of interpretation. We first turn to pragmatic typologies of *inferred meaning* before engaging with neo-Gricean and post-Gricean approaches to *explicit meaning*.

2 Pragmatic typologies of inferred meaning and interpretation in international law

There have been various attempts in pragmatics to systematise pragmatic interpretation. Our goal here is not to revisit these different approaches, but to rely on a particularly comprehensive typology recently developed by Mira Ariel to

---

[12] Iran-United States Claims Tribunal, *United States and Federal Reserve Bank of New York v. Iran and Bank Markazi*, Decision of 19 December 2000, Case A 28, [2000] Iran-US Claims Tribunal Reports, Vol. 36, 5, para 58.

show how such a systematic account of pragmatic interpretations could be used to describe and analyse interpretation in international law. We thus first present core aspects of Ariel's theoretical approach, namely her approach to explicit meaning and then her typology. Subsequently, we turn to the individual types of pragmatic interpretations and pair them with international law examples, to show the usefulness of such a typology for international law.

2.1 Mira Ariel's perspective on explicit meaning

Ariel's point of departure for her typology is a perspective influenced by Relevance Theory. *Pragmatic inferencing* is essential for understanding a speaker's communicative intentions; and there is simultaneous decoding of explicit messages and inferring of implicit messages, with resulting products that possess a distinct *discoursal status*. Put simply, a discoursal status corresponds to what people would say a sentence or text is "about," if asked, in the sense of what the speaker can be taken to have communicated or meant. Discoursal statuses are graded, with the gradation corresponding to how prominent these meanings are in discourse (Ariel 2016: 28). Explicit messages tend to be more prominent with regard to their discoursal status than implicit messages (Ariel 2016: 1). Before we can engage in more detail with her typology's foundations, we first need a primer on Ariel's perspective on explicit meaning.

In this regard, one must differentiate Gricean as well as neo-Gricean – presented below in section 4 with a focus on Horn – from post-Gricean pragmatics, specifically from Relevance Theory. Namely, the latter challenges the former's assumption that any pragmatic inference is by definition some type of implicature. Instead, the relevance-theoretic account distinguishes between *implicatures* and *inferences* constituting part of the *explicature* expressed by the utterance, which is the intended assumption explicitly communicated. Explicatures are based on the linguistic meaning decoded from the communicator's utterance. Since decoded meaning is underdetermined, the latter only serves as a starting point for developing the full proposition (or propositional form) the communicator intends to convey. This process relies heavily on inferencing, yet inferences in the service of enriching or completing the underdetermined decoded content in order to arrive at a truth-evaluable proposition are considered explicitly conveyed (Ariel 2008: 21).

These inferences are considered explicitly conveyed because context-independent decoding of the semantic elements expressed by the communicator yields only a minimal proposition, which is normally "uninformative, irrelevant, and sometimes truistic or patently false" (Carston 2006: 639). Such a minimal proposition would, therefore, be very rarely communicated, that is, speaker-meant or, in other words, would not yield sufficient cognitive effects to be relevant. Take the example of "Mary has no brain," which, except for in very particular medical contexts, would have to be pragmatically enriched to "Mary has no high-functioning brain" to satisfy expectations of relevance. Implicatures can thus only be drawn on the basis of the enriched proposition (Carston 2006: 639). Relevance Theory concludes that the Gricean conception of "what is said" may not be necessary in utterance comprehension, because semantically encoded meaning falls short of "what is said," while the content of explicatures goes well beyond "what is said," requiring pragmatic contributions for its recovery, just as implicatures do (Carston 2006: 636).

Ariel's perspective adds another nuance. According to Sperber and Wilson, the founders of Relevance Theory, the same cognitive principle of relevance responsible for deriving implicatures is also responsible for the pragmatic enrichment of the decoded meaning resulting in explicatures (Ariel 2008: 22). The difference between the relevance-theoretic conception of explicature and Ariel's explicature (described below in section 3.3.1) is that Ariel appears to focus on what she calls explicated inferences, that is, only the pragmatic inferences that are integrated into (Ariel & Mauri 2018: 946) or participate in forming the explicatures, because to her "there is no sense in comparing implicatures with all components of the explicature, because explicatures include encoded meanings as well" (Ariel 2008: 21).

What differentiates these two complementary perspectives from the Gricean position is that there is no explicature, because any inferred interpretation must be some kind of implicature (Ariel 2008: 78). What post-Griceans – and neo-Griceans — agree upon is the original Gricean position that implicatures do not contribute to or affect truth-conditions (Ariel 2008: 79, 83). According to Relevance Theory, since some inferences do, however, contribute to truth-conditions, they are therefore labelled explicatures.

2.2 Mira Ariel's typology

Let us now return to our example of "Can you pass me the salt?" to explain the difference between explicit and implicit messages and set out Ariel's approach to discoursal status. Inference is involved in the retrieval of both the explicit and implicit messages. While all pragmatic inferences are in principle cancellable, not all can be easily denied by the speaker (Sternau 2014). In the context of "Can you pass me the salt?", a speaker could add "In the sense of 'can you reach it,'" thereby clarifying that she is asking a question about the addressee's ability to reach the salt shaker, perhaps because the salt shaker is nearly out of reach for the addressee. Such inferences have a less prominent discoursal status. By comparison, it is more difficult for a speaker to deny inferences based on the explicit content of a message. These inferences are thus more prominent. In our example of "Can you pass me the salt?", the addressee must infer on the basis of the explicit element of "pass" how the salt shaker is to be transmitted, by hand or perhaps by foot (which would correspond to a "pass" in the context of football/soccer). It would appear very difficult for the speaker to cancel, or to deny, inferences based on this explicit content, in the sense that the addressee should not pragmatically (i.e. contextually) interpret "pass" at all. According to Relevance Theory, a purely semantic interpretation — that is, an interpretation without contextual information — would normally provide a result that is too underspecified to achieve relevance (the actually most relevant interpretation being that "pass" is to be understood as "by hand").

We can rephrase these ideas in the language of international law. Cases of implicit messages are those in which something is "not in the text of a treaty," as international lawyers would typically phrase it, but an interpreting agent nonetheless suggests a particular interpretation to apply a norm to a specific set of facts. The interpreting agent thereby ascribes a pragmatic interpretation to the relevant norm of the treaty at issue, considering it to be the speaker's (i.e. the treaty parties') intention. Thus, the interpreting agent takes the position that this is the correct interpretation. It is easier to defend this position when the pragmatic interpretation at issue has a more rather than less prominent discoursal status. The advantage from the perspective of international law is that the relationship between the interpretation and the treaty norm can be explained with higher precision than by merely suggesting that something is or is not "in the text," as often happens in international law.

Ariel's approach to the prominence of pragmatic interpretations is focused on comparing pragmatic interpretations with one another. For our purposes, if this idea is to be transposed to the context of international law and interpretation, another facet becomes of interest; namely, that categorising interpretations can help us to establish the extent to which an interpretation is far-fetched. Legally speaking, this in turn makes it possible to conclude that a stronger legal justification for an interpretation — based on elements other than the ordinary meaning — may be necessary in a particular context in which an interpretation has a less prominent discoursal status, and vice versa.

2.3 Applying the typology to interpretation in international law

Ariel distinguishes six types of pragmatic interpretations. In the present section, we follow her order which ranks these types by their strength (in the sense of the propensity to count as the speaker's relevant contribution) (Ariel 2016: 28). Due to constraints of space, not all details of Ariel's discussion can be taken up in the following overview. Our focus lies instead on pairing Ariel's types of interpretations and her non-legal examples with international law cases. In this manner, we want to demonstrate that types of pragmatic interpretations "fit" typical situations of interpretation in international law, and can help to offer a precise description of what happens during a process of interpretation. In what follows, we set out each of Ariel's types of pragmatic interpretation, explain the test Ariel suggests based on her own and others' work, in order to identify each type in relation to her non-legal example (Searle 1978: 207; Searle 1980: 221; Sperber & Wilson 1995; Ariel 2002; Jaszczolt 2005; Ariel 2004) and add a legal example. Ariel mainly uses excerpts from a newspaper article on so-called honour killings for her examples.

2.3.1 Explicature

As explained above, explicature or explicated inference designates the fact that an addressee must develop, i.e. pragmatically enrich or adjust, certain explicit elements of the utterance at issue in a situation of interpretation, in order to correctly interpret the speaker's intended meaning. In Ariel's typology, explicatures have the strongest and most prominent discoursal status (Ariel 2016: 6). Let us look at an example, and then develop the elements that are shown in italics.

> My son said that *she* wasn't the last *one*. We're waiting for the next *one*.

> The speaker's son said that *Busaina Abu Ghanem* wasn't the last *female murder victim in the family*. We're waiting for the next *female murder victim in the family*.

As a test to identify explicatures, Ariel recommends the "that-is" test. One adds a "that is (to say)" clause to spell out the explicated inference. If the pragmatic interpretation is an explicature, a correct utterance is the result, as our example shows.

> The speaker's son said that she, *that is (to say) Busaina Abu Ghanem* wasn't the last one, *that is (to say) the last female murder victim in the family*. We're waiting for the next one, *that is (to say) the next female murder victim in the family*.

In the case of explicatures, pragmatic inferences are limited to adjustments of the proposition that is expressed; they are not available as separate interpretations, and form part of a single meaning layer with the linguistic, or semantic, meaning (Ariel 2016: 11). The application of the test shows that there is no revision of the original example, only a development — i.e. a pragmatic enrichment or adjustment — of its elements.

In international law, we can test the usefulness of this category of pragmatic interpretation with the example of the case of *Maritime Delimitation and Territorial Questions between Qatar and Bahrain*. The parties disagreed on how exactly the parties could bring their dispute to, i.e. "seize," the International Court of Justice on the basis of minutes of talks between the two parties. Could one party seize the Court unilaterally, or did both parties have to agree to seize the Court? The minutes stated that "[o]nce th[e] period [of good offices of the King of Saudi Arabia] has elapsed, the two parties may submit the matter to the International Court of Justice [. . .]." Asked for its interpretation, the International Court of Justice held that the first part of the statement necessarily implied that there was a right by one party to seize the Court on its own after the mentioned period. Otherwise this phrase would have no effect, since the parties could always jointly seize the Court at any point in time if they so wished.[13]

Linguistically speaking, we can explain the Court's interpretation more explicitly by using the concept of an explicature. The Court has to interpret "the two parties" to decide whether either of the parties can seize the Court on its own. The utterance does not provide the answer to the question through its encoded meaning; semantically, it is possible to interpret "the two parties" as meaning "the two parties jointly" or as "either of the parties." If we apply the "that-is" test, we find that the Court's interpretation is an explicature and — importantly for international lawyers — it therefore has a relatively prominent discoursal status:

> Original: Once that period has lapsed, *the two parties* may submit the matter to the International Court of Justice.

> The Court's interpretation: *Either one of the parties* may submit the matter to the International Court of Justice.

> Test: Once that period has lapsed, the two parties, *that is (to say) either one of the parties*, may submit the matter to the International Court of Justice.

---

13 *Maritime Delimitation and Territorial Questions between Qatar and Bahrain, Jurisdiction and Admissibility*, Judgment, 15 February 1995, ICJ Reports (1995) 6, para 35.

With its argument that the clause needs to be given effect, the Court implicitly seems to be aware of the fact that the present explicature is part of a single meaning layer of the utterance, one which must be developed to interpret the speaker's intended meaning.

2.3.2 Strong implicature

Strong implicature has a strong, but still weaker, discoursal status than explicature: it would fail the "that-is" test (Ariel 2016: 30). In other words, strong implicatures are not explicitly, but only implicitly communicated: they are not developments of the bare linguistic meaning, but are independent of it and have separate truth conditions (Ariel 2016: 20). Strong implicature designates the situation in which a speaker says one thing (a first tier) but intends another (a second tier); a certain interpretation is not expressed directly, but the speaker intends that the directly communicated meaning is ultimately to be replaced with the intended interpretation (Ariel 2016: 4), as in the following example.

> R_1: And John Doe, who is a company director, pretends to know that the balance sheet is going to be good so he starts buying.
> S: OK that's a criminal offence.
> R_2: Eh . . .
> S: It's a bit of a criminal offence.
> R_3: So he has a mother-in-law.
> S: For this you go to jail.

R does not only directly communicate that John Doe has a mother-in-law in the example, but also strongly implicates that he would illegally buy shares under her name. Ariel (2016: 20) suggests an extended "replacement" test to identify such cases.

> The speaker *literally* said that John Doe has a mother-in-law, *but he actually indirectly conveyed that* John Doe would illegally buy shares under his mother-in-law's name.

In law, interpreters may act similarly when they replace what has been written down in a treaty, that is, the ordinary meaning, through legal interpretation. For example, in the *Les Verts* case, the Court of Justice of the European Union[14] had to decide whether a judicial remedy could be brought before it against acts of the

---

[14] For the present purposes, the distinction between international and European Union law is not relevant and can be left aside.

European Parliament having legal effects. The then relevant legal provision stated that "[t]he Court of Justice shall review the lawfulness of acts other than recommendations or opinions [i.e. acts not having binding legal effects] of the Council and the Commission." Even though the provision does not mention the Parliament, the Court famously argued that there had to be such a remedy because the Community (the predecessor of the Union) was a "community based on the rule of law" in which neither Member States nor institutions could avoid a review of their acts against the benchmark of the Treaty.[15] The Court mainly argued that the Parliament had only been left out of the provision because it initially had been given only weak powers, and that in the parallel European Coal and Steel Community Treaty, the Parliament had been given more powers from the beginning, and its acts could therefore also be attacked in Court. Leaving aside the legal arguments, linguistically speaking, Ariel's typology allows for a categorization of the case as one of strong implicature. Namely, the Court argues that the treaty text is to be read as strongly implicating that acts of "all the institutions handing down acts with legal effects towards third parties" can be reviewed by the Court.

> Original: The Court of Justice shall review the lawfulness of acts other than recommendations or opinions of the Council and the Commission.
>
> The Court's interpretation: The Court of Justice shall review the lawfulness of acts other than recommendations or opinions of the Council, the Commission *and the Parliament*.
>
> Test: The treaty *literally* stated that the Court of Justice shall review the lawfulness of acts other than recommendations or opinions of the Council and the Commission, *but it actually indirectly conveyed that* the Court of Justice shall review all acts of all EU institutions with binding legal effects towards third parties.

The advantage of describing the Court's approach in these terms is that it becomes clearer that the Court is not simply inventing elements that were not laid down in the treaty, but is replacing existing elements with — in its view — other, more plausible elements, or giving these elements a more pragmatically plausible interpretation. This does not mean that, legally speaking, the Court is necessarily "right." But it focuses the discussion on whether the replacement is justified, rather than on the "invention" of new elements.

---

[15] Case 294/83, *Parti écologiste « Les Verts » v. European Parliament* (EU :C :1986 :166), para 23–25.

2.3.3 Provisional explicature

Provisional explicature also designates the situation in which a speaker says one thing (a first tier) but intends another (a second tier). Contrary to the otherwise similar case of strong implicature, the speaker directly expresses the intended interpretation in the case of provisional explicature, but ultimately removes her commitment to the interpretation (Ariel 2016: 4). Provisional explicatures are, for example, cases of irony or rhetorical questions (Clark and Gerrig 1984: 121). Ariel presents the following example of irony and suggests the application of the previously mentioned "replacement" test.

> Women are murdered here and nobody cares. *Family honour!*
>
> The speaker *literally* said that women are murdered here and nobody cares and that that was family honour, *but she actually indirectly conveyed that* that was not at all family honour.

The difference to a strong implicature in such cases of irony is that the addressee is expected to hold on to the literal first tier to arrive at the intended implicated interpretation and to appreciate the difference between the two representations (Ariel 2016: 24).

Irony is not a particularly frequent phenomenon in legal interpretation, but some rare instances can be identified. Take the Paris Declaration Respecting Maritime Law of 1856, which concerned a disagreement over the legal status of privateering. The declaration prohibited privateering in the following terms: "Privateering is, and remains, abolished."[16] In a typical legal prohibition, it is unnecessary to state that a certain behaviour "remains" abolished because prohibitions always apply to the future; the use of "remains" therefore points to the past, namely to the fact that, in the view of the speaker, privateering has already been abolished in the past. It is thereby emphasised that the declaration merely restates an existing legal state of affairs. Seen in its historical context, the declaration is a message from the European states to the United States whose constitution provided Congress with the authority to name privateers.[17] The United States justified this practice on the basis of their own navy being comparatively weak (particularly in comparison with the then powerful navies of several European states), and needing reinforcement of its ranks by private merchant ships (Cooperstein 2009: 245–246). Put in terms of international law, contrary to the United States' view, the European states wanted to claim that

---

[16] Article 1, Declaration Respecting Maritime Law, Paris, 16 April 1856, LXI *British State Papers* (1856) 155–158.
[17] Article I section 8 of the Constitution of the United States.

customary international law already prohibited privateering before the Paris Declaration, and intended to reinforce this view by means of the Declaration. The addressee — the United States — ought to appreciate the gap between the following two representations and understand that the European states strongly disagreed with the United States' position (denying the existence of a customary international law prohibition or at least of being bound by this prohibition). Let us now put this in Ariel's terms.

> Original: The treaty stated that privateering is, *and remains*, abolished.

> Test: The treaty *literally* stated that privateering is, and remains, abolished, *but it actually indirectly conveyed that* privateering has not been universally abolished in the past, yet it should have been.

This explanation is arguably a fuller rendition of the intended interpretation process of the utterance (i.e. the Declaration) than to simply comment that it appears bizarre, wrong or unfounded to claim that privateering "remains" abolished. Of course, it also assumes that interpreters have the materials or knowledge at hand to arrive at this fuller rendition, or, put simply, that the United States representatives are able to "get" the irony.

2.3.4 Particularised conversational implicatures

Particularised conversational implicatures are additional implicatures conveyed by explicit content which are perceived as being separate from the content of the explicature on which they are based, and which have their own truth conditions (Ariel 2016: 11). They have a weaker discoursal status than provisional explicatures (Ariel 2016: 28). Take the following example from Ariel.

> Last Saturday night, Busaina Abu Ghanem was murdered, *the tenth female victim in the family*.

The reporter mentions the fact that she is the tenth victim for a reason, namely, to convey something additional. This could be, for example:

> There is something terribly wrong with this family.

Ariel suggests an "indirect-addition" test to identify such particularised conversational implicatures.

> The speaker said that last Saturday night, Busaina Abu Ghanem was murdered, the tenth female victim in the family, *and in addition she indirectly conveyed that* there is something terribly wrong with this family.

In law, the recent *Wightman* judgement in the Brexit context can be understood as involving a particularised conversational implicature which was self-constructed by the Court of Justice of the European Union. Put simply, the relevant provision (Article 50 of the Treaty on European Union) provides that any Member State may decide to withdraw from the Union in accordance with its own constitutional requirements (paragraph 1). A Member State that decides to withdraw "shall notify the European Council of its intention." Then, the parties are to negotiate an agreement on the withdrawal (paragraph 2). The question raised by the case was whether a Member State which has given notification of its intention to withdraw (like at that time the United Kingdom) could withdraw this notification unilaterally; a question clearly not answered by the text of the provision. The question is important because the possibility of a unilateral withdrawal of the notification could open the doors to abuse by the Member State in question. It was therefore suggested by some that there was a requirement of having the consent of the other Member States — in the form of a unanimous vote by the European Council.

The Court noted the silence of the provision on the question, but also that an "intention" was, by its nature, neither definitive nor irrevocable. Focusing on the broader context of the provision, the Court held that the provision as a whole pursued two objectives. First, it provided for a "sovereign right" of a Member State to withdraw on the basis of its own constitutional requirements; second, it provided a procedure for withdrawal. The parallel provision on accession to the European Union made it clear in the Court's view that the Union was composed of states having "freely and voluntarily" committed themselves to common values. If a state could not be forced to join the Union, it could also in the Court's view not be forced to withdraw from it against its will. The Court thus decided that the unilateral revocation of the notification was possible as a sovereign decision of a Member State.[18]

Put in linguistic terms, we can represent the bare provision and the Court's interpretation of it as two individual utterances, neither of which is a pragmatic interpretation of the other in the sense of an explicature.

> Original: The treaty stated that a Member State that decides to withdraw shall notify the European Council of its intention.

> The Court's interpretation: A Member State is free to decide to unilaterally revoke such a notification.

---

18 Case C-621/18, *Wightman a.o.* (EU:C:2018:999), paras 50–65.

However, the Court presented its reasoning differently, namely as a particularised conversational implicature, which creates a stronger link between the interpretation and the norm. In particular, the Court did so by drawing from context, adding the fact that a Member State had just as much a sovereign right to decide to withdraw from as to accede to the European Union. Let us try to put this in more linguistic terms.

> Elements added by the Court: The treaty stated that a Member State *is free to decide to join and leave* the European Union *as its sovereign right in accordance with its own constitutional requirements*.

> Test: The treaty further stated that a Member State that decides to withdraw shall notify the European Council of its intention, *and in addition it indirectly conveyed that* a Member State is free to decide to unilaterally revoke such a notification.

The possibility of the unilateral revocation is separate from the content of the explicature on which it is based, and it has its own truth conditions. The utterance as presented by the Court actively participates in shaping the implicated conclusion. Put simply, the Court tries to make its interpretation appear more logical than it would be based solely on the treaty text.

2.3.5 Background assumptions

Background assumptions are implicit aspects of the communication that the speaker makes because they are part of her world knowledge. At the same time, the speaker does not assume responsibility for communicating them. They are merely assumed (Ariel 2016: 15) or taken for granted and thus have a weaker discoursal status than particularised conversational implicatures (Ariel 2016: 28), which the speaker wishes to convey (Ariel 2016: 13). Ariel uses the following example.

> The victim's house has no customary mourner's booth and no visitors appear.

> Background assumption: It is customary to have a mourner's booth where visitors come to pay their respect to the dead.

She suggests a "circumstantial-report" test to identify background assumptions.

> The speaker said that the victim's house has no customary mourner's booth and no visitors appear. *She intended the addressee to take into consideration the fact that* it is customary to have a mourner's booth where visitors come to pay their respect to the dead.

The speaker thus entertains background assumptions and intends the addressee to access them, but such assumptions are not directly or indirectly part of the message communicated by the speaker (Ariel 2016: 17).

To provide an example from international law, it is prohibited in international humanitarian law to declare that no quarter will be given. The relevant norm, Article 23 (d) of the Annex to the Hague Convention on the Laws and Customs of War (IV),[19] provides that, among other things, it is prohibited "[t]o *declare* that no quarter will be given." It is typically argued in the legal doctrine that if it is prohibited to order or threaten that no quarter shall be given, it must also be prohibited to carry out such an order or threat during a military operation.[20] The linguistic "link" between such a suggested interpretation and the original norm text can be better explained by the concept of background assumptions, as can be shown with the help of the "circumstantial-report" test.

> Original: The treaty stated that it is prohibited to declare that no quarter will be given.

> Typical interpretation: It is prohibited to deny quarter.

> Test: The treaty stated that it is prohibited to declare that no quarter will be given. *It intended that the addressee take into consideration the fact that* it is prohibited to deny quarter.

This linguistic description shows that there is a link — admittedly weak, but nonetheless existing — between the norm and its interpretation.

2.3.6 Truth-compatible inferences

Truth-compatible inferences are implicit inferences that a speaker is likely to endorse, given the content of her utterance or assumptions that will be seen as compatible with the speaker's utterance should the assumption be true in reality (Ariel 2016: 24).[21] Such inferences are merely potentially derived but not speaker intended. They thus have the weakest discoursal status (Ariel 2016: 28) of Ariel's six types of pragmatic interpretations. Take the following example used by Ariel.

---

[19] Annex to the Hague Convention respecting the Laws and Customs of War (IV), Oct. 18, 1907, 187 Consolidated Treaties Series 227 (emphasis added).
[20] For an overview of the doctrine, see Henckaerts and Doswald-Beck (2005: 162).
[21] In the sense that the speaker is seen as not having precluded such an assumption.

> *In the wake of the murder* of Busaina Abu Ghanem last weekend, activists say the police must do more to intervene.

In Ariel's example, the journalist reports only that after the last murder there are complaints from activists, as she finds this fact relevant. However, there have been complaints by activists before, and the journalist in all likelihood knows this fact. Ariel suggests a very weak "compatibility" test to identify truth-compatible inferences.

> The journalist said that in the wake of the murder of Busaina Abu Ghanem last weekend, activists say the police must do more to intervene. *Her utterance is compatible with a state of affairs in which* the activists spoke up before.

For a legal example, we can turn to the *Bosnia Genocide (Merits)* case.[22] The International Court of Justice was asked to decide whether the Genocide Convention[23] created an obligation for the parties to the Convention not to commit genocide. Article I of the Convention provides that the parties confirm that genocide is a crime under international law and that they undertake to prevent and punish it. In its decision, the Court held that the "actual terms" of the Convention did not contain an obligation on the states; Article I did not require states to refrain from committing genocide *expressis verbis*. But, according to the Court, taking into account the purpose of the Convention, the "effect" of Article I was nonetheless to prohibit states from committing genocide. This was due to the fact that the Convention categorised genocide as a crime under international law, logically meaning that states could not undertake such an act after having agreed to this categorisation. Moreover, states had to prevent persons or groups not directly under their control from committing genocide under the Convention. It would be "paradoxical" in this light, according to the Court, if states were not forbidden to commit such acts themselves by their own organs, bodies or persons over which they had sufficient control. Therefore, in the Court's view, the obligation to prevent genocide "necessarily implie[d]" the prohibition to commit genocide.[24]

---

22 *Application of the Convention on the Prevention and Punishment of the Crime of Genocide (Bosnia and Herzegovina v. Serbia and Montenegro) (Bosnia Genocide), Merits*, Judgment, 26 February 2007, ICJ Reports (2007) 23. It should be noted that our assessment is not a disagreement with the actual outcome of that case, with which we wholly agree. Our contention is rather that a more intense linguistic engagement could have led, in our view, to a more convincing reasoning underpinning the decision.
23 Convention on the Prevention and Punishment of the Crime of Genocide, 9 December 1948, United Nations Treaty Series, vol. 78, p. 277.
24 Para 166 of the judgment.

Put in linguistic terms, this is a truth-compatible inference. We can use Ariel's "compatibility" test to show this.

> Original: The treaty stated that the parties agree that genocide is a crime under international law and undertake to prevent and punish it.
>
> The Court's interpretation: The parties themselves are obliged not to commit genocide.
>
> Test: The treaty stated that the parties agree that genocide is a crime under international law and undertake to prevent and punish it. *Its utterance is compatible with a state of affairs in which* the parties themselves are obliged not to commit genocide.

There is thus a link between the utterance and its interpretation, even if it is rather weak. This means that, in turn, a powerful legal justification will be required to make the suggested interpretation acceptable from the point of view of international law.

2.4 Interim conclusion

We have thus far presented Ariel's typology and have brought it together with examples taken from law. The discussion shows that linguistic categorisations — in particular with regard to the discourse status that an interpretation has — are often more precise than the explanation for interpretations given by the relevant interpreting agents, who are often courts. Legally speaking, the conclusions presented in these cases are, of course, tenable. If, however, lawyers are to take their own rules of interpretation seriously, and these rules typically prescribe to explain interpretations in light of the "ordinary meaning" of treaty texts, typologies of pragmatic interpretations can be of help to make the legal reasoning more transparent and more compliant with this ordinary meaning requirement. This is not to suggest that the rules of (international) law are changed thereby; quite the contrary, they can be followed with more precision by relying on refined linguistic categorisations. Namely, the less prominent the discoursal status of a particular interpretation is, the stronger a legal justification will be needed and vice versa.

3 Neo-Gricean vs. post-Gricean perspectives on explicature and legal interpretation

Having examined Ariel's typology of pragmatic interpretation types and applied it to legal examples, we can now return to one of these types, namely explicature, to illustrate the neo- and post-Gricean perspectives on this type of pragmatic inference and the respective accounts as an example of legal interpretation, to provisionally assess their usefulness to the latter. Having set out a general post-Gricean perspective on explicit inference or explicature (and Ariel's particular approach to explicated inferences) above (section 3.1), we can now contrast this perspective with a neo-Gricean one, relying on Horn's views for this purpose.

3.1 Horn's views on explicature

As a neo-Gricean, Horn holds that propositions enriched or strengthened by pragmatic inferencing cannot be explicatures, because they are not explicit in the sense of not directly expressed. In other words, what is explicated under Relevance Theory is, in Horn's view, an indirectly communicated proposition (see Ariel 2008: 81; Horn 2006). In his view, the pragmatically strengthened proposition can only be implicitly – but not explicitly – communicated, for the reason that it involves cancellable pragmatic inferences. He therefore objects to the notion of explicature (Horn 2005: 193). This objection is to be understood against the backdrop that Horn's orientation is linguistic, concentrating on "the most systematic and least context-sensitive aspects of pragmatics" (Carston 2005: 306). Unlike Relevance Theory, Horn does not intend to provide a cognitive account of utterance processing. Whereas Relevance Theory argues that there is no need for Gricean maxims and that relevance alone is sufficient, which is not considered a maxim or communicative principle but a basic feature of cognitive processes (Birner 2013: 92), Horn reduces the Gricean maxims to two principles: the Q Principle (say as much as you can, given R) and the R Principle (say no more than you must, given Q) (Horn 1984: 12–13; Horn 2009: 14), The Q Principle minimises the speaker's effort, while the R Principle minimises the hearer's effort (Carston 2006: 306). These two principles serve cooperation in the Gricean sense, with the tension between them potentially giving rise to implicatures (Birner 2013: 85).

The interaction between the two principles is guided by Horn's *division of pragmatic labor* (2005: 196): "an unmarked utterance licenses an R-inference to the unmarked situation, whereas a marked utterance licenses a Q-inference to the effect that the unmarked situation does not hold. (An 'unmarked' expression is in general the default, usual, or expected expression, whereas a "marked" expression is non-default, less common, or relatively unexpected)" (Birner 2013: 80). Take the example: "Mary's jacket is light red." The colour "pink" is a subtype or shade of "light red," but not all shades of light red are pink. Pink, however, is the default or prototypical variety of light red. The use of the expression "light red" thus licences a Q-based inference that Mary's jacket is not pink because the Q Principle dictates that "pink" be used if her jacket could be described this way. A specific context may, however, override such inferences (Birner 2013: 81). It has to be noted that Horn's implicatures do not arise automatically nor by default (Horn 2009: 22) but "in a default context," i.e. in the absence of special circumstances (Horn 2009: 22) or specific contexts, which means that Horn endorses only a "'weak' defaultism" (Horn 2009: 23). As an example of such specific contexts which may override this weak defaultism, Birner provides the following: "If I come in from mowing the lawn on a 95-degree day and utter 'I need a drink,' my hearer is more likely to infer that I need liquid refreshment, and that lemonade would serve the purpose"; this non-default context thus overrides the unmarked case in which "to need a drink is to need an alcoholic drink" (Birner 2013: 81).

It is noteworthy that there are cases in which it is not clear whether the Q or the R Principle applies, even when there is no specific context. Consider the example: "There is a vehicle in the driveway." Is the word "vehicle" an unmarked expression, which thus licences an R-based implicature to an unmarked situation in which the correct interpretation would probably be a motor car; or, given that there is a salient semantic entailment or Horn scale <vehicle, car>, does the choice of the semantically weaker expression "vehicle" licence a Q-based implicature that it is not a car? From the perspective of Relevance Theory, a properly developed concept of communicative relevance would play a key role in accounting for why, in certain contexts, it is correct to choose one interpretation over the other (cf. Carston 2013: 16).

3.2 A legal example

Psycholinguistic experiments may prove helpful in determining which of these accounts of inference drawing, that is, the neo-Gricean predictions on implicatures in default contexts or the post-Gricean predictions on contextually inferred interpretations, are correct (cf. Zufferey et al. 2019: 143, 165). To recapitulate, according to the neo-Gricean (as well as the Gricean) position there is no post-Gricean explicature because any inferred interpretation is, by definition, implicit and must therefore be some kind of implicature (Ariel 2008: 78). In the preceding discussion of explicature (section 3.3.1), the focus was on the post-Gricean view that an addressee must develop, i.e. pragmatically enrich or adjust, certain explicit elements of the utterance, which means that the resulting explicature is part of what is explicitly communicated (and thus contributes to the truth-conditions of the explicitly expressed utterance). In other words, this is why explicature passes Ariel's "that-is" test. The focus will now be on the question of whether the drawing of certain legal inferences can be better modelled by the post-Gricean explicature (or explicated inference), i.e. the pragmatic (or contextual) enrichment of the decoded meaning based on the cognitive principle of relevance, or Horn's neo-Gricean Q and R Principles (or maxims), which, based on the division of pragmatic labor, result in diametrically opposed implicatures in a default context.

As a preliminary measure, let us look at an example taken from law, which will be examined through the prism of Ariel's typology (cf. Ariel 2016) as well as from a Hornian perspective.[25] The example is taken from international law,[26] more precisely from EU law Directive 2009/103/EC, which prescribes that EU Member States must ensure that civil liability "in respect of the use of vehicles normally based in its territory is covered by insurance."[27] The Directive defines vehicles as "any motor vehicle intended for travel on land and propelled by mechanical power, but not running on rails, and any trailer, whether or not coupled" (Article 1 pt. 1). As the case law shows, the European Court of Justice had to deal with various scenarios, such as military vehicles or damages caused by stationary vehicles or vehicles being used for purposes other than transport.[28] The Court had to decide whether these instances fall under the definition of "use of vehicles." When examined against Ariel's typology described in section 3, the Court's interpretation of "use of vehicles" seems to allow an explicature in many

---

25 Many thanks to Professor Laurence Horn for his constructive criticism of the authors' presentation at the 2019 ILLA Conference and his ordinary language example taken up below, which lent us the impetus to reflect on this problem. Many thanks also to Professor Didier Maillat for discussing the points raised by Professor Horn with one of the authors and for making helpful suggestions.
26 Once again, the distinction between EU law and international law is not relevant for the present purposes, as the focus lies on interpretive phenomena that would arguably be treated similarly in classic international law cases.
27 Article 3 of the Directive, Official Journal of the European Union 2009, L 263, 11–31.
28 Case C-334/16, *Núñez Torreiro* (EU:C:2017:1007); Case C-648/17, *BTA Baltic Insurance Company* (EU:C:2018 :917); Case C-100/18, *Línea Directa Aseguradora* (EU:C:2019:517); Case C-514/16, *de Andrade* (EU:C:2017:908).

cases. As discussed above, Ariel's explicature or, rather, explicit meaning, appears to be compatible with or complementary to the relevance-theoretic one.

To offer an example, let us apply Ariel's "that-is" test (by adding a "that is (to say)" clause) — which explicature passes — to one of the above-mentioned cases the Court had to decide. In order to apply the test, let us somewhat simplify the wording of the directive, that is: "The Directive states that civil liability in respect of the use of vehicles must be covered by insurance." The case in question concerned a tractor which had reversed with a trailer in the courtyard of a farm and struck a ladder on which a worker had climbed, causing him to fall.[29]

> The Court's interpretation: "The Directive states that civil liability in respect of the use of vehicles, *that is (to say)* a tractor reversing with a trailer in the courtyard of a farm, must be covered by insurance."

This test is compatible with the relevance-theoretic notion of explicature, that is, the construction of an *ad hoc* concept (cf. Carston 2013: 12), here "use of VEHICLE*." Explicature does not mean that everything goes or that a vehicle can be anything, since explicatures are based on the cognitive principle of relevance. In other scenarios, the Courts used various arguments on the normal function of a vehicle as serving as a means of transport to exclude the explicature or the creation of an *ad hoc* concept, for example, in the case of a stationary tractor with the engine running to drive a spray pump for herbicide.[30]

Let us now look at the example from a neo-Gricean perspective. Horn would argue that the type of pragmatic inference presented in the example was from a *hypernym* (vehicle) to a *hyponym* (tractor). Hypernym and hyponym form a scale which allows for an implicature from hyponym to hypernym, but not the other way around, as presented in the example (from vehicle to tractor). Horn's reasoning seems correct when applied to ordinary language. Consider the example orally provided to the authors by Horn: "This dog, that is to say an animal" (implicature from hyponym to hypernym) feels natural, while "this animal, that is to say a dog" (implicature from hypernym to hyponym) does not. It appears that — as far as ordinary conversation is concerned — within Horn's theoretical framework, an implicature to a subset/subclass/hyponym of a category might only be licenced if the latter is a stereotypical exemplar of the category. Otherwise, as the dog example shows, such an implicature may not be licenced in ordinary conversation.

---

[29] Case C-162/13, *Vnuk* (EU:C:2014:2146).
[30] Case C-514/16, *de Andrade* (EU:C:2017:908).

To return to the tractor example, there is, as indicated by Horn, a salient Horn scale <vehicle, tractor>. Following Horn's approach, the use of the expression "vehicle," which is the semantically weaker expression, licences a Q-based implicature that it is not a tractor because if it were, the semantically stronger expression should have been used. As already mentioned, Horn's Q-based and R-based inferences are defeasible inferences that will not hold in all contexts. To give an example of how the Q-based implicature could be contextually overridden, one could think of a legal definition provided for vehicles including an exhaustive list mentioning tractors as examples of vehicles. What is, however, missing from Horn's account are the above-mentioned rules of interpretation. They do not have to be explicitly stated, such as in the example of the legal definition, to apply to (or as context for) legal interpretation. One may take up relevance-theorist Carston's suggestion that neo-Gricean principles, i.e. the Q and R Principles, are very similar to rules of interpretation in the sense that they are sometimes employed explicitly and sometimes implicitly (2013: 16). The content of the two principles (or maxims), i.e. "say as much as you can, given R" (Q Principle) and "say no more than you must, given Q" (R Principle), would not normally be explicitly stated in communication. Similarly, the rules (or maxims) of interpretation are not always explicitly mentioned. Take the case of the ordinary meaning rule (or maxim): it is always applied, yet only mentioned when the parties' dispute specifically relates to it. Returning to our example, it appears to be the case that the specific legal context, which includes, for example, the legal principle or rule "like cases need to be treated alike" (see below), overrides the Q-based implicature, and even allows for a "counterintuitive" one (i.e. 'the use of vehicles, that is to say a tractor').

To put this more precisely: while the illustrative ordinary language example ("This dog, that is to say an animal" feels natural, while "this animal, that is to say a dog" does not) appears perfectly suitable to ordinary conversation, a case of legal interpretation will not normally deal with a specific occurrence of an object or person as expressed by the demonstrative "this [dog]." Instead, the law must always start from a general legal term, which may often be a hypernym, for example, "animal" or "vehicle." This is due to the fact that terms in national and international legal texts have to encompass a wide range of possible occurrences, that is, they do not normally refer to a specific occurrence of an object or person. Whenever a legal term has to be defined by a court, the court has to narrow it down — or broaden it — keeping in mind generalisable characteristics. The substantive legal principle that the court applies can be described as "like cases need to be treated alike." It thus appears to be the case that the specific legal context or principle/rule of interpretation overrides the Q-based implicature — although it appears questionable that the context provided in the example would not count as a default context in legal interpretation — and even allows for a pragmatic inference that appears counterintuitive from the perspective of ordinary communication.

From the relevance-theoretic perspective, an inference in the form of an explicature or *ad hoc* concept avoids this problem, because it is context-sensitive and does not depend on default contexts. In order to construct an explicature, an addressee will access manifest assumptions in his or her cognitive environment (see section 2.1), which includes background knowledge of relevant information about sociocultural dimensions (cf. Foster-Cohen 2004: 289, 294, 300), such as a legal expert's knowledge of the rules of interpretation (cf. Smolka & Pirker 2016: 26). Knowledge of the rules of interpretation can thus be described as a set of highly accessible assumptions or propositions which are mentally represented (Carston 2013: 29) in the mind of the legal expert.

In the legal example, the explicature thus appears to be better able to capture legal reasoning in the sense that an *ad hoc* concept can involve a broadening or a narrowing process, or even both of these processes, with the goal of making the "use of vehicles" more informative and more specific, that is, relevant in the context of legal interpretation. In other words, instead of a uni-directional inference on the basis of a Horn scale, which can only be overridden by non-default contexts — which the context provided in the example does not seem to be — explicature appears to allow for a bi-directional (or rather multi-directional) pragmatic inference.[31]

One may counter that the mentioned dog example still holds without the demonstrative "this," for example, "the/a dog, that is to say an animal" versus "the/an animal, that is to say a dog." The aforementioned legal perspective, however, still applies: not only do terms in (international) legal texts not normally refer to a specific occurrence of an object or person, but one may also argue that these terms are so generic that their referent is missing, that is, they are not applicable to any set of individuals. Referents are only assigned when a court has to define such terms, which explains why — although it appears counterintuitive from the perspective of ordinary communication — a pragmatic inference from hypernym to (a non-stereotypical) hyponym is licenced in legal interpretation.

---

31 A legal example of an *ad hoc* concept involving broadening (following initial narrowing) can be found in Pirker and Smolka (2017: 256): to establish the ordinary meaning of "land" in the context of the question of where prisoners of war can be held captive, a legal interpreter narrowed the semantic content of "land" to *firm ground with housing structures on it, which serve as shelter/protection from the elements* and then broadened this *ad hoc* concept to include "land" that is adjacent to it, even if it is technically located on top of water, i.e. *ships*.

3.3 Interim conclusion

While experimental research is needed to confirm the provisional conclusion drawn from this example, it appears as if the post-Gricean account of ordinary utterance interpretation can, "perhaps with certain provisos and/or modifications [to account for the rules of interpretation]" (Carston 2013: 32), be applied to legal interpretation and make better predictions about the drawing of legal inferences than the neo-Gricean account. What can also be concluded from the example, although this preliminary result obviously needs further consideration, is that Carston's claim that the rules of interpretation "are instantiations of Gricean maxims" (2013: 17) as well as of "neo-Gricean maxims" (2013: 18)[32] might need to be nuanced. In fact, in legal interpretation, it is not only the case that legal interpreters have to choose between conflicting rules of interpretation (cf. Carston 2013: 18) or maxims, but neither of the aforementioned neo-Gricean principles or maxims could model or predict the legal inference in the example and — unless the rules of interpretation count as a (non-default) context that can override these principles — even yielded a legally unusable result.

4 Conclusion and outlook

The present chapter has presented two possible approaches to refine legal thinking on interpretation by relying on pragmatics. The normative framework of (treaty) interpretation in international law chosen here as an example shows that semantic and pragmatic concepts can be transferred to legal interpretation. As a first useful approach, pragmatic typologies of inferred meaning, such as Ariel's, offer more precise linguistic categorisations than the explanations for interpretations typically provided by interpreting agents in international law, such as courts. Arguably, lawyers could benefit from such increased precision without challenging thereby the normative framework within which they operate; concepts such as "ordinary meaning" actually require them to justify their interpretations as precisely as possible.

Second, a closer look at the notion of explicature from a neo-Gricean and a post-Gricean perspective allows the preliminary conclusion that the latter may be more suitable to make predictions about the drawing of legal inferences. Neo-Gricean maxims appear even to yield a legally unusable result, which at least

---

[32] It must be noted that Carston's claim appears to be drawn from an analysis of a mixture of Levinson's and Horn's, not only from Horn's neo-Gricean account (cf. Carston 2013: 13 ff.).

partly contradicts the views of scholars such as Carston. Notably, legal interpretation normally means going from the general to the specific, or inferring more specific terms on the basis of more general terms, which may not be compatible with Horn's notion of a default context. The question is to what extent should such a default context only apply to ordinary conversation, or where should the line be drawn between ordinary everyday and specialised (legal) communication and/or context.

In addition to what has been suggested so far, there are also methodological opportunities that open up for (international) law if the mentioned suggestions to rely on pragmatic knowledge are accepted. International law has recently started to engage in experimentation (Chilton & Tingley 2013; Shereshevsky & Noah 2018; Dunoff & Pollack 2018; van Aaken 2019; Marceddu & Ortolani 2020). It thereby follows in the footsteps of experimental pragmatics (Noveck 2018). Nothing seems to stand in the way of future projects relying on experimental methodology to gain new insights into the role of pragmatics in legal interpretation. The operation in international law of typologies of pragmatic interpretations[33] or the accuracy of the preliminary conclusions drawn in the present chapter on neo- and post-Gricean approaches to the drawing of pragmatic inferences in legal contexts could therefore be tested. Such experiments could yield highly relevant insights both for pragmatics and for (international) law.

References

Ariel, Mira. 2002. Privileged interactional interpretations. *Journal of Pragmatics* 34 (8). 1003–1044.
Ariel, Mira. 2004. Most. *Language* 80 (4). 658–706.
Ariel, Mira. 2008. *Pragmatics and grammar*. Cambridge, UK: Cambridge University Press.
Ariel, Mira. 2016. Revisiting the typology of pragmatic interpretations. *Intercultural Pragmatics* 13 (1). 1–36.
Ariel, Mira & Caterina Mauri. 2018. Why use or? *Linguistics* 56 (5). 939–993.
Austin, John L. 1962. *How to do things with words*. Oxford: Oxford University Press.
Aust, Helmut Philipp, & Georg Nolte (eds.). 2016. *The interpretation of international law by domestic courts: Uniformity, diversity, convergence*. Oxford: Oxford University Press.
Birner, Betty. 2013. *Introduction to pragmatics*. Malden, MA: Wiley-Blackwell.
Bix, Brian. 2012. Legal interpretation and the philosophy of language. In Peter Tiersma & Lawrence Solan (eds.), *The Oxford handbook of language and law*, 145–155. Oxford: Oxford University Press.

---

[33] See the *International law, linguistics and experimentation* (IntLLEx) project: https://www3.unifr.ch/ius/euroinstitut/de/forschung/forschungsprojekte.html

Bjorge, Eirik. 2014. *The evolutionary interpretation of treaties*. Oxford: Oxford University Press.
Börjesson, Kristin. 2014. *The semantics-pragmatics controversy*. Berlin & Boston: De Gruyter.
Carston, Robyn. 2005. Relevance Theory, Grice, and the neo-Griceans: A response to Laurence Horn's 'Current issues in neo-Gricean pragmatics.' *Intercultural Pragmatics* 2 (3). 303–319.
Carston, Robyn. 2006. Relevance Theory and the saying/implicating distinction. In Laurence Horn & Gregory Ward (eds.), *The handbook of pragmatics*, 633–656. Oxford: Blackwell.
Carston, Robyn. 2013. Legal texts and canons of construction: A view from current pragmatic theory. In Michael Freeman & Fiona Smith (eds.), *Law and language*, 8–33. Oxford: Oxford University Press.
Chasapis Tassinis, Orfeas. 2020. Customary international law: Interpretation from beginning to end. *European Journal of International Law* 31 (1). 235–267.
Chilton, Adam & Dustin Tingley. 2013. Why the study of international law needs experiments. *Columbia Journal of Transnational Law* 52 (1). 173–237.
Clark, Herbert H. & Richard J. Gerrig. 1984. On the pretense theory of irony. *Journal of Experimental Psychology* 113. 121–126.
Cooperstein, Theodore M. 2009. Letters of marque and reprisal: The constitutional law and practice of privateering. *Journal of Maritime Law and Commerce* 40 (2). 221–260.
Dörr, Oliver. 2018. Article 31: General rule of interpretation. In Oliver Dörr & Kirsten Schmalenbach (eds.), *Vienna convention on the law of treaties: A commentary*, 559–616. Berlin & Heidelberg: Springer.
Dunoff, Jeffrey & Mark Pollack. 2018. Experimenting with international law. *European Journal of International Law* 28 (4). 1317–1340.
Foster-Cohen, Susan. 2004. Relevance Theory, Action Theory and second language communication strategies. *Second Language Research* 20 (3). 289–302.
Fyfe, Shannon. 2017. Tracking hate speech acts as incitement to genocide in international criminal law. *Leiden Journal of International Law* 30 (2). 523–548.
Grice, Paul. 1975. Logic and conversation. In Peter Cole & Jerry Morgan (eds.), *Syntax and Semantics 3: Pragmatics*, 41–58. New York: Academic Press.
Henckaerts, Jean-Marie & Louise Doswald-Beck. 2005. *Customary international humanitarian law*. Cambridge, UK: Cambridge University Press.
Horn, Laurence. 1984. Toward a new taxonomy for pragmatic inference: Q-based and R-based implicature. In Deborah Schiffrin (ed.), *Meaning, form, and use in context: linguistic applications*, 11–42. Washington, DC: Georgetown University Press.
Horn, Laurence. 2005. Current issues in neo-Gricean pragmatics. *Intercultural Pragmatics* 2 (2). 191–204.
Horn, Laurence. 2006. The border wars: A neo-Gricean perspective. In Klaus von Heusinger & Ken Turner (eds.), *Where semantics meets pragmatics*, 21–48. Amsterdam: Elsevier.
Horn, Laurence. 2009. WJ-40: Implicature, truth, and meaning. *International Review of Pragmatics* 1 (1). 3–34.
Jaszczolt, Kasia. 2005. *Default semantics: Foundations of a compositional theory of acts of communication*. Oxford: Oxford University Press.
Klabbers, Jan. 2017. *International law*, 2nd edn. Cambridge, UK: Cambridge University Press.
Klingler, Joseph, Yuri Parkhomenko & Constantinos Salonidis (eds.). 2019. *Between the lines of the Vienna convention? Canons and other principles of interpretation in public international law*. Berlin: Kluwer Academic.

Linderfalk, Ulf. 2007. *On the interpretation of treaties: The modern international law as expressed in the 1969 Vienna convention on the law of treaties*. Dordrecht: Springer.
Linderfalk, Ulf. 2013. All the things that you can do with Jus Cogens: A pragmatic approach to legal language. *German Yearbook of International Law* 56. 351–383.
Linderfalk, Ulf. 2018. What are the functions of the general principles? Good faith and international legal pragmatics. *Zeitschrift für ausländisches öffentliches Recht und Völkerrecht* 78 (1). 1–31.
Macagno, Fabrizio, Douglas Walton & Giovanni Sartor. 2018. Pragmatic maxims and presumptions in legal interpretation. *Law and Philosophy* 37 (1). 69–115.
Marceddu, Maria Laura & Pietro Ortolani. 2020. What is wrong with investment arbitration? Evidence from a set of behavioural experiments. *European Journal of International Law* 31 (2). 405–428.
McNair, Arnold Duncan. 1961. *The law of treaties*. Oxford: Clarendon Press.
Merkouris, Panos. 2017. Interpreting the customary rules on interpretation. *International Community Law Review* 19 (1). 126–155.
Moeschler, Jacques. 2009. Pragmatics, propositional and non-propositional Effects: Can a theory of utterance interpretation account for emotions in verbal communication? *Social Science Information* 48 (3). 447–463.
Moeschler, Jacques. 2010. Is pragmatics of discourse possible? In Alessandro Capone (ed.), *Perspectives on language, use and pragmatics: A volume in memory of Sorin Stati*, 217–241. Munich: Lincom Europa.
Moeschler, Jacques & Antoine Auchlin. 2009. *Introduction à la linguistique contemporaine*, 3rd edn. Paris: Armand Colin.
Noveck, Ira. 2018. *Experimental pragmatics: The making of a cognitive science*. Cambridge, UK: Cambridge University Press.
Pirker, Benedikt & Jennifer Smolka. 2017. Making interpretation more explicit: International law and pragmatics. *Nordic Journal of International Law* 86 (2). 228–266.
Reboul, Anne & Jacques Moeschler. 1998. *La pragmatique aujourd'hui: Une nouvelle science de la communication*. Paris: Editions du Seuil.
Regan, Donald H. 2017. Sources of international trade law: Understanding what the Vienna convention says about identifying and using 'sources for treaty interpretation.' In Samantha Besson & Jean D'Aspremont (eds.), *The Oxford handbook on the sources of international law*, 1047–1065. Oxford: Oxford University Press.
Sbisà, Marina. 2006. After Grice: Neo- and post-perspectives. *Journal of Pragmatics* 38 (12). 2223–2234.
Sbisà, Marina & Ken Turner. 2013. Introduction. In Marina Sbisà & Ken Turner (eds.), *Pragmatics of speech actions*, 1–24. Berlin & Boston: De Gruyter.
Searle, John. 1969. *Speech acts: An essay in the philosophy of language*. Cambridge, UK: Cambridge University Press.
Searle, John. 1978. Literal meaning. *Erkenntnis* 13. 207–224.
Searle, John. 1980. The background of meaning. In John Searle, Ferenc Kiefer & Manfred Bierwisch (eds.), *Speech act theory and pragmatics*, 221–232. Boston: Reidel.
Shereshevsky, Yahli & Tom Noah. 2018. Does exposure to preparatory work affect treaty interpretation? An experimental study on international law students and experts. *European Journal of International Law* 28 (4). 1287–1316.
Slocum, Brian G. 2015. *Ordinary meaning: A theory of the most fundamental principle of legal interpretation*. Chicago: Chicago University Press.

Slocum, Brian G. & Jarrod Wong. Forthcoming. The Vienna Convention and the ordinary meaning of international law. *Yale Journal of International Law* 46 (2).
Smolka, Jennifer & Benedikt Pirker. 2016. International law and pragmatics: An account of interpretation in international law. *International Journal of Language & Law* 5. 1–38.
Sorel, Jean-Marc & Valérie Boré Eveno. 2011. Article 31: Convention of 1969. In Olivier. Corten & Pierre Klein (eds.), *The Vienna conventions on the law of treaties: A commentary*, 804–840. Oxford: Oxford University Press.
Sperber, Dan & Deirdre Wilson. 1995. *Relevance: Communication and cognition*, 2nd edn. Oxford: Blackwell.
Sternau, Marit. 2014. *Levels of interpretation: Linguistic meaning and inferences*. Tel Aviv: Tel Aviv University dissertation.
Van Aaken, Anne. 2019. Experimental insights for international legal theory. *European Journal of International Law* 30 (4). 1237–1262.
Wilson, Deirdre. 2003. Relevance Theory and lexical pragmatics. *Italian Journal of Linguistics* 15 (2). 273–291.
Wilson, Deirdre & Dan Sperber. 2006. Relevance Theory. In Laurence Horn & Gregory Ward (eds.), *The handbook of pragmatics*, 607–632. Oxford: Blackwell.
Zufferey, Sandrine & Jacques Moeschler. 2012. *Initiation à l'étude du sens*. Auxerre: Sciences Humaines Éditions.
Zufferey, Sandrine, Jacques Moeschler & Anne Reboul. 2019. *Implicatures*. Cambridge, UK: Cambridge University Press.

Svetlana V. Vlasenko
Temporal meanings in legal translation: English-Russian lacunas and associated semantic uncertainties

> The ultimate validation of a translation can never be
> a purely linguistic undertaking.
> –Eugene A. Nida

Introduction

Today, Legal Translation Studies is being shaped to address multiple issues and challenging problems of ever-increasing complexity, across a wide range of issues. These may include, but are not limited to: legal translation contexts, legal genres, the collision of different legal constructs, principles, and institutions; interaction within multilingual jurisdictions; multilingualism in parliamentary practices and high-profile summits; interpreting at cross-border litigations, forensic translation, and many others.

Quite a few scholars, from either the legal profession or from outside of it, have acknowledged that legal translation is both technical and complicated. Over the past quarter century, sites for legal translation have expanded. In these settings, law and language interact and entwine in a variety of intricate ways. Focused interest from a considerable diversity of perspectives has resulted in ongoing research by legal scholars from different fields of law, forensic linguists, legal translators practicing nationally and internationally, translation studies scholars, and legal linguists. This avalanche of publications now amounts to a bulky corpus comprising Ramos (2019; 2021); Robertson (2016); Giltrow and Stein (2017); Gotti and Williams (2010); Cheng, Sin and Wagner (2016); Olsen, Lorz and Stein (2008; 2009); Tiersma and Solan (2012); Pozzo and Jacometti (2006); Šarčević (2000), among others.

Research problem and research tools

This chapter advocates the view that legal languages are complex, insofar as they serve conceptually heterogenous legal frameworks of domestic (national) laws.

Svetlana V. Vlasenko, Moscow State Linguistic University

https://doi.org/10.1515/9783110720969-008

Legal translation is therefore a challenging task, which involves not only linguistic and cross-linguistic competences, but also legal knowledge in respective fields of the law. Research into time and temporality expressed in legal contexts cross-linguistically reveals a complex interaction of universal concepts across the temporal continuum, while at the same time identifying certain dissimilarities that result in lacunas in legal Russian when it is translated into legal English.

It is generally understood that temporality, and time itself, are primordial concepts. Therefore, the universal nature of the experience of time is not in doubt. What can be argued, however, is that there are multiple linguistic realizations of temporality, and distinct ways of experiencing time, across cultures and languages. Culture- and language-specific manifestations of temporal meanings present a widely relevant research object. Undoubtedly, it would be valuable to inventory temporal constructs and their manifestations in the language of the legal professions, but this chapter does not aim to accomplish that task. Rather, it aims to show a set of English and Russian lexis, with culture- and language-specific temporal meanings, operating in legal contexts.

The allocation of temporal constructs along the time continuum makes them a challenging object of research, since no action or omission, a *bona fide* act or wrongdoing, abuse or grievance, malpractices, nor any other situations of misconduct, can be either described or alleged without reference to time. Hence, the temporal coordinates of any event, situation, condition, state of affairs, or circumstance constitute essential knowledge for any legal case. All legal facts or events must therefore accommodate relevant temporal coordinates, to facilitate essential legal decisions.

As for the research tools used in this chapter, the concept of time is treated in linguistic terms, with a particular focus on translation tools and the relevant techniques employed by legal translators when handling low-equivalent texts. These common tools and techniques include: contextual and/or situational analysis, bridging inference, analogical inference, as well as extensional interpretation and definition methods for describing the meanings of technical concepts with reference to their encyclopaedic or dictionary definitions.

1 Evolving global contexts as the metacontext for legal translation and interpreting

At present, researchers continue to explore the use of artificial intelligence for modelling and simulating translation processes. However, in spite of the wide availability of machine translation and welcomed improvements in its quality, the evolution of contexts due to human-induced innovations still outpaces all extant translation software. Countless technical issues requiring broad expert knowledge on the part of translators and interpreters, as well as sound knowledge of respective terminologies, constitute modern contexts for international talks, where the summiteers may be top politicians, parliamentarians, key experts, national and international media, and other major stakeholders. The language of the legal profession in national jurisdictions serves national legal systems, which are evolving in unison with these legal systems and international law as they cope with international legal challenges. The requirements for translation or interpreting for international milieus today exceed by far those of the past, regardless of the availability of electronic dictionaries, translation software, and translation webtools.

Compared with past agendas, the current agendas discussed by statesmen and key decision-makers globally via translators/interpreters address many more technical issues. This is unsurprising, since globalization has left no opportunity for countries to escape from the root causes of political, economic, environmental, and social challenges conspicuous worldwide. Today, agenda-setting in any one country involves not only politicians, but also legislators and economic policymakers, financiers and investors, national strategists, and local governments. The agendas to be tackled in the short-, mid- or long-term perspective vary within a wide range of problem areas, such as human-modified habitats, environmental pollution and endangered species; freshwater insufficiency and nuclear power facilities' maintenance and security; food safety management and nutrition insecurity; vaccine races for pandemics and appropriate use of biotechnologies; human trafficking and illicit trafficking of goods and services; novel IT installations going through legal loopholes in regulatory instruments; malware detection and cyberwarfare, cybercriminals and cyber policing; digital forensics; corruption and influence peddling at upper echelons of power; cosmic space garbage and cosmic law on space ownership, as well as international terrorism with global, regional, and national implications. These urgent, difficult matters, which have scarcely been settled nationally, let alone internationally, call on government officials to monitor, control and address them in collaboration, aided by trained interpreters and translators.

2 Would *time* translate?

The ways that temporal meanings are manifested in legal texts can alter legal relations of contracting parties or affect the construing of legal events, facts, and circumstances essential for adjudication in the law of procedure. Along with that, they may have profound effects on comprehending the rule of law or may cause disputes in noticeably different understandings of freshly drafted legal novellas. Issues associated with conveying temporal meanings are indeed widespread problem-areas in legal translation. Translating from English into Russian can be a precarious task, because the basic phrases and expressions denoting time-related constructs sometimes do not translate. This is not due to a lack of ideas in Russian about time. Rather, the difficulty is due to culture-specific usages which mismatch in tongues belonging to different language families. Wierzbicka (1997) delineated a large set of universal human concepts based on empirical studies in many languages, predictably called *universals*. The citation below outlines Wierzbicka's fundamental assumptions about language:

> We cut nature up, organize it into concepts, and ascribe significances as we do, largely because we are parties to an agreement to organize it in this way – an agreement that holds throughout our speech community and is codified in the patterns of our language.
>
> (1997: 8)

Among the concepts delineated, Wierzbicka compiled a table of universals with "time" as a category of human experience and cognition. This category comprises the following set of conceptual primes: *when (time), now, after, before, a long time, a short time, for some time* (1996: 97; 2002: 407). In view of this, there is every reason to expect every language to have concepts of time and temporality in its lexical repertoire which overlap, coincide, and share many conceptual similarities.

The two subsections below contain several examples contrasting simple semantics with more complicated manifestations used in both English and Russian to convey more or less the same temporal meanings in legal contexts.

3 Meanings of temporal constructions

This research on English-Russian legal translation focuses on the semantics and the pragmatic determinants of a range of English temporal constructions with long-established usage: *at all times*; *as the case may be / whatever the case may be*; *from time to time*; *lapse of time*, along with other words expressing temporality. These constructions, single words, and word combinations, employed in English and Russian legal languages to draw temporal coordinates of legally significant facts or events, are analysed from a semantic-pragmatic perspective. Translation correspondences are analysed to exemplify the challenges associated with cross-linguistic conveyance of temporal meanings.

Along with studying semantic properties and their pragmatic determinants, event-timing other technical words are explored, such as *sutki* and *dekada*. These words' meanings are predominantly determined text-externally through detailing specific contexts of their use, contexts which include legal regulations.

Meanings of temporal constructions usually aim to specify relations of *simultaneity*, *precedence*, or *sequence*, or to emphasize timed relations, such as time-framing, timelining, or continuity or discontinuity. For example:
– the meaning of localizing, or specifying, on the temporal continuum: *the day before yesterday*; *now*; *in a three-week period*; *two years ago*; *until recently*; *since the late 1960s*;
– the meaning of additional temporal referencing: **the then**-*presidential campaign advisor*; **the ex**-*wife's legal counsel*.

Many temporal modifiers – *late, recent, early, past* – are used to show the timing of the legally significant features of events, facts, or conditions involved in qualifying legal relations. If *early* marriages are addressed as a problem of the abuse of children's rights, the modifier *early* does not indicate to Russian lawyers whether a 7-year-old child or a 15-year-old teen is meant, because it can refer to either. Depending on what age-range is meant, English terms would differ: the former case would be called a *child marriage*, the latter -an *early marriage* The same applies to Russian wordings, if and when the age is known: *detskije braki* 'child marriages' and *rannije braki* 'early marriages' (Child marriages 2013; Looking ahead towards 2030).[1] Where no age is indicated, the modifier *early* would prevail.

3.1 Abstract time and its encoding

Temporal meanings may hamper a translation process when some deep-rooted nuances entwine with other conceptual primes in the national worldview, thus yielding culture- and language-specific meanings. Structurally, temporal constructions can be single-word constructions, such as *always, once, sometimes, occasionally*, or can be multi-word constructions, such as *at any time*; *lapse of time*. Longer constructions like *as the case may be / whatever the case may be*,

---

[1] Child marriages 2013: Child marriages: 39 000 every day / World Health Organization. Report. https://www.who.int/mediacentre/news/releases/2013/child_marriage_20130307/en/ (accessed 15 November 2020); Looking ahead towards 2030: Eliminating child marriage through a decade of action/ UNICEF. September 2020. https://data.unicef.org/resources/looking-ahead-towards-2030-eliminating-child-marriage-through-a-decade-of-action/ (accessed 15 November 2020).

from time to time, at all times are abundant and can be found across many legal texts, irrespective of genre or text type. Due to their wide distribution in legal English, these temporal constructions rightly merit the category of linguistic universals.

It is noteworthy that definitions of certain technical legal terms which entered American law dictionaries over 150 years ago contain the construction *at all times*. This mere fact attests to the durability of this temporal construction in the language of the legal profession. Given its importance in the language of US law, definitions of legal concepts containing this wording call for rigorous analysis. Two examples below have been selected for analysis from *Black's law dictionary* (2009). Each of the following entries contains the construction "at all times," which we have emphasized in boldface:

- *joint custody*. (1870) An arrangement by which both parents share the responsibility for and authority over the child **at all times**, although one parent may exercise primary physical custody. (Garner 2009: 442)
- *inherently dangerous*. (1887) (Of an activity or thing) requiring special precautions **at all times** to avoid injury; dangerous *per se*. (Ibid.: 451)

Communicating temporal meanings from English into Russian reveals the English form's rigidity and a clear reluctance of Russian to respond identically. The following example is convincing in showing that the construction *at all times*, which implies abstract, borderless time, just would not translate:
– We reached an agreement with them to cooperate fully at all times. – My zaklučili s nimi dogovor o tesnom i dolgovremennom sotrudničestve (ABBYY 2014).

The English phrase *to cooperate fully at all times* corresponds to the Russian (back-translated into English) [signed an agreement on] *close and long-term cooperation*, which are merely approximately equivalent in meaning.

Quite obviously, being a set construction in one language does not guarantee that same status for that construction in the other language, specifically where abstract time is implied. Since translation always presupposes two or more languages in contact, translation relying on set constructions in one language can be challenged in the other.

3.2 Legal technicalities associated with temporal constructions

As temporal space presents a universal concept, there are many ready-made translation correspondences, even in languages as distantly related as English and Russian. Here are a few examples:

- *na moment* [at the moment of / whereon / wherefrom]
- *v moment* [at the very moment [when]] or [at [exactly] the instance of / when]
- *v den' / na datu* [as of the date] or [as at the date of / when / whereon]
- *v period / za period* [over the period of] or [within the period of]

It is, however, not so much *moment* that is technical, as is the precise sequencing of timed events and the implications in the event of any breach of this prescribed sequence.

One expression frequently used in contract law, *at the commencement of* [the lease] or, its variant, *at the inception of the lease* [term], can be translated into Russian in the following way: *na načalo sroka [arendy]*. There is nothing technical about either phrase, which indicates only the starting point for the leasehold to be fixed in a lease deed and/or other relevant document enabling the leasehold. This reference to the "commencement of the lease term" may matter as a condition implying certain timed payment in the legally binding documents. Therefore, the *commencement of the lease* or the *commencement of the lease term* or the *inception of the lease* should be understood as complete synonyms, and be marked as such, in law dictionaries. This marking is lacking in law dictionaries.

The following two examples illustrate the technical nature of conditional clauses incorporating time-chained modifications in legal relations, where references to timing or consecutively chained events matter. Both examples are taken from the Russian explanatory comments to regulations on exercising taxation practices.

> - Dlya celey otraženiya v buchgalterskoj otčetnosti finansovyje vloženiya v zavisimosti ot sroka obraščeniya (pogašeniya) podrazdelyayutsya na kratkosročnyje i dolgosročnyje. Perevod dolgosročnoj zadolžennosty v kratkosročnuyu v časti vydannych zaymov proizvoditsya **v moment, kogda do okončaniya sroka pogašeniya ostajetsya 365 dney**

[With a view to entering records into accounting books, financial investments shall be split up into short-term and long-term based on maturity. Reclassifying the long-term debt into short-term as regards loans granted shall be made **at the date wherefrom three hundred and sixty-five (365) days are left for full redemption.**]

An alternate, more literal translation of the emphasized conditional construction, with its built-in temporal component, may read as follows: "at the moment when 365 days are left till the end of the repayment period."

The next example, below, is remarkable in that it is entirely based on the comprehension of rigidly chronological temporality of several consecutive events, whose priorities are allocated by relevant regulations and whose breach might have material implications (such as partially lost revenues).

> – *Esly **na moment perechoda prava sobstvennosty** po dogovoram postavky tovarov na vnutrennyj rynok, usloviyamy kotorych predusmatrivaetsya vozmožnost' izmeneniya ceny realizaciy po sravneniyu s cenoj, opredelennoj **na datu perechoda prava sobstvennosty**, vyručka priznaetsya v buchgalterskom učete po cene, soglasovannoj v dogovore **na datu perechoda prava sobstvennosty**.*
>
> [If **as at the date of transferring the proprietary right** under the contractual provisions for inbound supply of goods whose terms and conditions envisage an option for altering the selling price against **the price fixed as at the date of transferring the proprietary right**, receipts shall be recognized in the books of accounts **at the prior price agreed upon in the contract as at the date of transferring the proprietary right**.]

4 The split of the temporal continuum

As an indispensable conceptual framework for everyone in any part of the globe, the temporal continuum is split evenly and/or unevenly into constituent elements. The complexity and importance of this split is reflected in prolific research on linguistic means of objectifying the time-space relationship (see, for instance, Filipović, Katarzyna 2012; Lewandowska-Tomaszczyk 2016). Many researchers point out that time-markings are language-specific (Apresjan 2012; Gladkova 2012). In other words, languages tend to use their own ways for objectivizing time and temporality along the temporal continuum. Time is relativized against this language or another language or against a group of languages. Therefore, what matters and what differs is merely "packing" or "wrapping" of time or timing into temporal units – its constitutive "bricks," which do not correspond across languages (Stepanov 1997: 236).

One English phrase from insurance law can serve as a good example: *during the 12-month period immediately preceding the events causing the claimed liability*. This phrase would be awkward for Russian lawyers, who are used to employees' salary/wage accrual being measured in months, in contrast with the per annum count used in countries of Anglo-American law. Any Russian lawyer would instantly convert *the 12-month period* into *one year [period]* without another thought. Months are used in Russian administrative law to measure relevant types of offence, but not criminal liability, which is primarily measured in years.

4.1 Units of timing: a *year*

Financiers at large, and within the accounting profession, make extensive use of linguistic expressions to record and report cashflows of all kinds, be it revenue, expenditure, investment, or loss. Expressions from financial reports, statistical records, payrolls, or pension plans cannot be directly attributed to legal language, however, since these activities have to be exercised in total compliance with statutory norms and respective legal regulations, they may appear in legal opinions as well. Several examples below show a mismatch of English conciseness against Russian verbosity:
- **Y/Y = year on year / year-on-year** – *v godovom isčislenyi* [(as) compared with the figures of the last year's corresponding period]; *po sravneniyu s predydyščim godom* [as compared against the corresponding figures (prices, etc.) in the previous year] (Faekov 2011(I): 740);
- **YOY = year-over-year** *v godovom isčislenyi* [computed over the year / calculated based on records over the year to compare against the previous reporting period]; *sravnivayemyj s predydyščim godom* [(as) compared against the last year's figures (prices, etc.) of the corresponding period] (Ibid.);
- **YTD = year-to-date** – *za poslednyj god* [over the last year]; *s načala goda do nastoyaščego momenta / do tekuščej daty* [(period of time) from the (calendar or fiscal) year start through the current date] (Ibid.).

4.2 Undated: a shared concept with unshared encoding

A dateless situation or event, one with no marked date, can occur anywhere, no matter the language spoken. However, languages differ in the way they encode such situations. Legal English marks it in a variety of ways, including abbreviations in capital or lower-case letters: **N.D. > ND > n.d. > *no date*, *dateless*, *undated***, while legal Russian uses just a couple of ways to denote any such situation: *bez daty* [dateless], *data ne ukazana* [undated] or *nedatirovanno* [no date has been marked].

Strictly speaking, none of the above abbreviated or full representations of time intervals form part of the legal language *per se*. However, they can be found in litigation papers and fall under the scrupulous examination of investigators handling bankruptcy or insolvency proceedings, corporate fraud, tax evasion schemes, and the like, if and when these might be caused by such things as accounting irregularities, false financial statements or false tax returns.

Hence, the existing *lexical lacunas*, or gaps in the lexical stock of the language, uncover the lack of corresponding equivalents in the target language,

Russian. This *lacuna* reveals that the concept does exist in both languages, but that each language in the pair studied here, English and Russian, instead uses its own distinct cluster of expressions and/or abbreviations to encode the shared concept, and that those expressions mismatch.

In general, a *lacuna* as a psycholinguistic construct correlates predominantly with the perception of the text to be translated; it is often prompted by new and/or unshared knowledge which the translator/interpreter may lack in the source language or in the target language. As such, semantic opacity or vagueness caused by certain objective discrepancies in the lexical stocks and/or structures of the contacting language pair, is a feature of lexical lacunas, which prompt the translator/interpreter to resort to bridging inference, or analogical inferencing, or otherwise to compensate for the lexical gap. This type of lacuna is typically a *lexical lacuna*. In other cases, when there is no semantically shared concept in the contacting pair of languages, let alone its naming, a lacuna is recognized as a *conceptual lacuna* (Vlasenko 2006: 46–47). Lacunas are effects of semantic inconsistencies, the lack of accurate correspondences between the contacting languages. The most challenging of these are considered *deep lacunas*, such as the case of the Russian concept *sutki* examined in section 5.2.1 below.

4.3 A *moment* and an *instant* as a *date*: universal and/or culture-specific naming of temporal units

As far as timing is concerned, Russian speech patterns contain clusters of phrases with references to *moments* as markers of *exact points in time* underscoring the particular timing of events, state of affairs, or actions, such as *at the moment, at this moment,* and *at the present moment, right after the moment.* Undoubtedly, this preference for *moments* does not in any way entail an exclusion of longer time periods – such as days, weeks, months – which are indeed marked in Russian. It is noteworthy that the expression containing the word *instant – year to the first instant –* can provide us with a convenient bridge to commenting on Russian temporal preferences and expressed meanings, while keeping in mind the universality of the experience of time.

Russian speakers prefer *moment* for marking a "particular time." The phrase *at the time of his arrest* would likely be translated into Russian as *at the moment of his arrest.* Indeed, a *moment* or an *instant* in a Russian contract can be conveyed as the "time" or a "date" when translated into English.

The source-language text (SLT) is a Russian contract for services showing inherent inclination to using a *moment* in legal texts:

Otkaz Subpodryadčika ot učastiya v meropriyatiy ne vlečet za soboj vozvrata Organizatorom- denežnych sredstv, perečislennych po nastoyaščemu Dogovoru **na moment napravlenniya sootvetstvuyueščgo uvedomleniya.** *Denežnye sredstva, perečislennyje po nastoyaščemu Dogovoru* **na moment napravlenniya uvedomleniya,** *podležat uderžaniyu Organizatorom v pokrytiye raschodov i kompensacyi za ponesjennye ubytki.* **S momenta rastorženiya nastoyaščego Dogovora** *Organizator vprave ispol'zovat' oborudovaniye, vydelennoye Subpodryadčiku soglasno usloviyam Dogovora, po svoyemu usmotreniyu.*

The target language text (TLT) options are English alternative translations:

(a) TLT-1: The Subcontractor's refusal to participate in the event shall entail no obligation to the Organizer to reimburse the funds transferred as per the provisions hereunder specified by the Subcontractor **at the time of sending the correspondent notification**. The funds transferred in line with the terms and conditions specified hereunder **at the time the correspondent notification is sent** shall be withdrawn by the Organizer to cover the costs inflicted and recover the loss suffered. **From the date the Contract is deemed terminated**, the Organizer shall be entitled to use the equipment, allotted to the Subcontractor under the provisions hereunder, at the Organizer's own discretion.

(b) TLT-2: Refusal by the Subcontractor to partake in the event shall imply no reimbursement by the Organizer of the funds transferred pursuant to the terms hereunder by the Subcontractor **at the date the due notification is sent**. The funds transferred pursuant to the terms hereunder **at the date of sending the due notification** shall be retained by the Organizer to cover the costs incurred and the loss inflicted. **Upon the termination of this Contract**, the Organizer shall enjoy the rights of using the equipment designated to the Subcontractor pursuant to the contractual terms and conditions hereunder indicated at the own sole discretion of the Organizer.

(c) TLT-3: The Subcontractor's refusal to be engaged in the event shall inflict no recompense to the Organizer of the funds transferred pursuant to the terms specified hereunder by the Subcontractor **at the time of notifying the latter**. The funds transferred as per the provisions hereof as **at the time of notifying the Organizer** shall be retained by the Organizer for the cost-recovery purposes, as well as for recompensing the lost suffered. **Once the Contract is terminated**, the Organizer shall be authorized to use all the equipment assigned to the Subcontractor recovering as per the terms and conditions specified hereunder at sole discretion of the Organizer.

The mere availability of several translation options through (a) TLT-1 to (c) TLT-3 indicates the variety of means available for encoding the same meanings, by employing more than one set of conceptual "wrappings." In this case, as shown, there are three alternative versions in English used to convey the technical Russian timing associated with the contractual language and respective terms. The three different versions, however, seem to coincide around *moment* in the SLT in all three technical wordings. There is reason to see *moment* as the nation-specific word used to frame even verbose technical passages of text denoting points in time.

Looked at together, the variety of options available in English for indicating the same point in time indicated by *moment* in Russian seems impressive:

(a) **at the time** of sending the correspondent notification; **at the time** the correspondent notification is sent; **from the date** the Contract is deemed terminated;
(b) **at the date** the due notification is sent; at the **date** of sending the due notification; **upon the termination** of this Contract;
(c) **at the time** of notifying the latter; at the **time** of notifying the Organizer; **once** the Contract is **terminated**.

Given the above examples, there is every reason to expect that the translation options analysed here can only approximate Russian legal meanings to the patterns of perception and speech of lawyers from other cultures, and this can be expected to hold true for practically all temporal dimensions that a legal instrument may convey.

5 Temporal meanings in legal texts as translation challenges

Apresjan (2012), in her thorough analysis of Russian temporal phrasemes, reveals the correlation of semantic and pragmatic factors that trigger the choice of interpretation for the following temporal phrasemes: *second, minute, hour, day, week, month,* and *year* (2012: 114–115). In so doing, Apresjan overlooks frequently used words such as a *moment,* an *instance,* and the distinctively Russian term *sutki.* Gladkova (2012), addressing the conceptualization of time in Russian, confirms the universal dimension of certain Russian denotations of time, including the most general – *vremja* [time], *sejčas* [now], *pora* [it is time], *teper'* [nowadays, currently], and *nynče* [today] – while also indicating their culture-specific and language-specific traits (2012: 167–169). Although Apresjan's and Gladkova's general approaches are well-reasoned and validated, some units typical of the Russian temporal continuum have been missed. Despite the exacting approaches offered, neither included *sutki,* a unit of the temporal continuum which has been indispensable to Russian mindsets for centuries.

5.1 Temporal dimensions of legal meanings

Legal meanings operate so that the legal relations conveyed in legal texts help "establish, alter, or terminate legal relations" (Tiersma 2001: 72). An event or activity, or a series of events or activities, representing either a realized intention or will of one party towards another, can be expected to fall within a certain domain of legal regulation and must include temporal coordinates as prerequisites of such legal relations. Given this, any legal text, no matter its genre, contains words referring to time or expressing temporal meanings.

5.2 Framing temporal concepts cross-linguistically

Temporal meanings can hamper translation, due to conventional beliefs in certain boundaries that various jurisdictions may have in their terms of temporal referencing, in particular with respect to temporal patterns rooted in perception of a day and night. That is to say, there are culturally specific applications of the 24-hour framework as a measurement of time that can pose challenges in translation. Nation-specific temporal assumptions and language-specific speech patterns conveying temporal meanings may also require that the translation strategy resort to versions which accord with current usages in the given legal culture. Temporal terminology therefore becomes an issue. Words denoting temporal properties add up to common sensitivities typical of legal languages. The following cases of *sutki* and *dekada* are useful for illuminating the mindsets behind the wording practices.

5.2.1 The case of a lexical lacuna: 24 hours vs *sutki*

The Russian word *sutki* stands in opposition to its respective stereotypes in the English language, wherein the temporal semantics of a common word *day* may be treated as limited or, on the contrary, unlimited in its temporal dimensions. The case of *sutki* is illustrative, because 24 hours as a temporal concept exists both in the English and Russian languages. However, the former uses the concept differently by naming this period one of the following ways: *all day long, (a)round the clock, day and night, all day and all night, twenty-four hours, a 24-hour period*, while the latter has just a single word for the concept, *sutki*. Russian-English dictionary definitions of *sutki* read as follows:

- twenty-four hours, a twenty-four-hour period (Oxford Russian 2000: 502)
- a time-measuring unit / a timekeeping unit equalling 24 hours; a continuous time span lasting 24 hours from one midnight to another (Yefremova 2005: 547, cited as in ABBYY 2014)

Given these definitions, differences between the representation of the 24-hour framework in English and Russian are classified below in Table 1. The table is intended to show that there is no conceptual difference in the interval/period. Rather, there are different ways of encoding the period, which give rise to those linguistic dissimilarities which, as we have said, are called lexical gaps, or lexical lacunas.

Table 1: A case of a lexical lacuna: representations of *24 hours* and *sutki* in English and Russian.

24 hours – sutki / kruglosutočno			
English		**Russian**	
Denoted meaning(s)	Time length (quantifiable unit of time – an *hour*)	Denoted meaning(s)	Time length (quantifiable unit of time – an *hour*)
all day long *(a)round the clock* *day and night* *all day and all night* *twenty-four hours* *a 24-hour period*	24 hours	**sutki** [all day long, (a)round the clock] **kruglosutočno** [daily; day and night, all day and all night; twenty-four-hour period] **dvadcat' četyre časa** [twenty-four hours]	24 hours

Remarkably enough, Russian speakers typically perceive *sutki* as an integral, uninterrupted period. Nonetheless, this does not preclude perceiving *sutki* as a period of time split into a day and a night. To this end, *sutki* is conceptualized from a variety of legal perspectives (see Figures 1–3 below) as a culturally specific way of experiencing time. Basically, scholars point out in their research on temporality that "one usually finds it difficult to slice the continuous flux of space-time into a series of events with clear-cut temporal boundaries" (Huang 2012: 38).

Figures 1 to 3 showcase the notion of *sutki* conceptualized via simple graphics as a temporal space unit and the experience of time across various legal contexts.

Figure 1: Conceptualization of *sutki* as a temporal whole: an integral, non-discrete, and continuous space, valid across certain legal fields and respective regulations.

Figure 1 formats *sutki* to simplify its conceptualized perception as temporal whole, i.e. uninterrupted temporal space. This concept of *sutki* applies across certain legal fields and is defined in relevant regulations, from calculating *per diem* during secondments or business trips, to establishing a period of confinement for a minor offence at detention facilities. The implied nature of such experience of time is text-external.

Figure 2. Conceptualization of *sutki* as a divided temporal unit adjustable to various legal contexts for regulation purposes.

Figure 2A: *Sutki* conceptualized as a temporal space incorporating two components – a day and a night – to be applied for regulatory purposes across relevant legal fields.

Figure 2A depicts *sutki* conceptualized as a temporal space combining two components – a day and a night – which can be adjusted to various legal contexts for regulation purposes to accord with specific legal relations and culturally specific experiences of time. Such conceptualization enables regulations and instruments, with a view to applying different legal regimes to certain professions and respective professional activities, to be exercised strictly within daytime hours or strictly within night-time hours, as envisaged by the legal instruments in question. Such activities may include long nonstop flights, long uninterrupted labour shifts (especially covering the period between midnight and

8 a.m.), medical emergency nonstop work carried on overnight, lifesaving and rescue operations, as well as many other instances of professional works and services singled out and differentiated in regulations underscoring the time interval within which such work has to be done or services rendered. Crossing the 00:00 threshold is treated as an extreme extension of labour hours and accrues additional compensation.

Such conceptualization of *sutki* is determined by pragmatic factors, and yields the text-external timing shown in Figure 2A.

Figure 2B: *Sutki* conceptualized as discrete temporal spaces, subject to the statutory day-night split for empowering rules, including *ad hoc* rules, to administer public order and/or contend with externalities.

Figure 2B presents *sutki* as discrete temporal spaces, subject to the statutory day-night split. This concept of *sutki* applies with a view to enabling the implementation of different legal regimes, which are in place to administer public order. These may include fixed hours for pre-schooling and schooling; defined timeframes for certain types of labour; timing restrictions for the physically challenged or exclusion of overtime hours at a workplace for certain citizenry groups; time restrictions for visiting the elderly at medical facilities; proactive responses to any breach of timing restrictions which keep people indoors within specified hours (typically at night), or outlawing any movements whatsoever during urgent save-and-rescue operations, states of emergency or wartime, and/or imposing penalties during curfew, wartime, or other such situations. Crossing those deadlines set by authorities entails the infliction of penalties or detention for offenders. Deadlines are set separately for each case of calamity and are treated as a temporary timing rule caused by externalities, calamities, or other such extreme conditions. Circumventing the deadlines officially fixed in *ad hoc* rules/regulations is treated as a minor offence or misdemeanour and has legal implications.

Restrictions imposed regarding the timing of public events at specified intervals reveal pragmatically determined comprehension of *sutki* as discrete temporal spaces. Also, it evidences the clear text-external nature of such conceptualization.

Temporal meanings in legal translation — 181

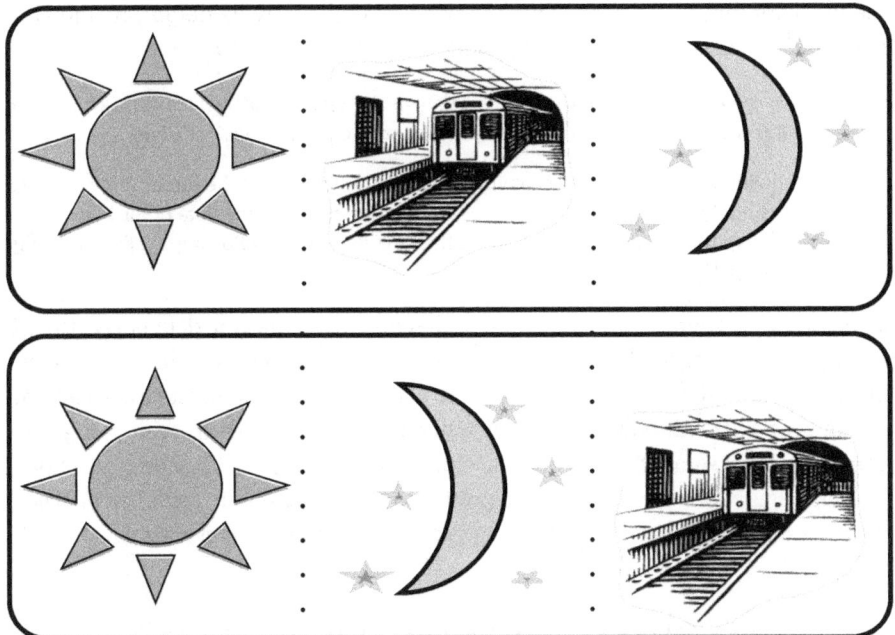

Figure 3: Generic conceptualizations of *sutki* as temporal spaces applicable to different fields of law regulating transport, education, medicine, etc. and burdened with timed arrangements.

Figure 3 features the conceptual perception of *sutki* as continuous temporal spaces, burdened with time-tabled arrangements and/or practices, which fall under various regulations. A 24-hour period is treated differently in various fields of law. Essentially, this period may begin at the birth of a stillborn baby or at the sudden death of an employee at the workplace and, as such, it may be indicated in relevant legal documents – birth or death certificates – as evidence for insurance coverage or the beginning of a coroner's inquiry. This concept of *sutki* is specifically important, as it is rarely tied to the 00:00 hour. It starts when an event in question actually occurs, and it accrues from that particular time – in Russian often called a "moment."

These conceptualizations of the Russian language- and culture-specific construct *sutki* have clearly been determined by pragmatic factors, and have thus yielded text-external timing – externalities, calamities, or other such emergencies (on text-external determinants of legal meanings see Vlasenko 2016a,b; 2017).

When do *sutki* begin and how long do they last? If the length of *sutki* is no issue, as it equals 24 hours, the beginning of that period can alter depending on the legal context in question. The following example may be helpful in showing that *sutki* can equally well be correlated precisely with 00:00.

- Lico sčitaetsja dostigšim vozrasta, s kotorogo nastupajet ugolovnaja otvetstvennost', ne v den' roždenyja, a po ego istečenyji, to est s nolja časov sledujuščich sutok.

[An individual shall be deemed having reached the legal age wherefrom (s)he shall be held criminally liable before the court of law starting not from the date of birth but at the expiration of this date, i.e. from 00:00 of the next day [sutki] following the date of birth]. (RF Supreme Court Resolution 2012)

In addition, vagueness in event time and states of affairs (for details see Huang 2012), where events or states of affairs depend on externalities, can be a factor complicating translation and the matching of equivalent correspondences. Therefore, readings of *sutki* will obviously be predominantly context-bound, even though the word encodes a semantically universal concept of "24 hours/all day around." Some Russian regulations name a daytime portion within the 24-hour period as *dnevnoje vremya sutok*, while naming a night-time portion as *nočnoje vremya sutok*.

5.2.2 Treating *sutki* as event-time determined text-externally

The construct *sutki* can be regarded as a concept perceived by law-enforcement agencies as a temporal unit of measurement, with its length/duration equalling 24 hours, but with its start-end properties varying if it accrues in situations of detention and confinement. In this case, inception and cessation of *sutki* are determined by statutory provisions (criminal law of procedure), as in the following excerpt from the Russian Federation Penal Procedure Code (UPK RF 2012):

- Ugolovno-processyal'nyj kodeks Rossijskoj Federacii ot 18.12.2001 N 174-FZ (red. ot 12.11.2012) (s izm. i dop., vstupayuščimy v silu s 21.11.2012). Statjya 128. Isčislenyje sroka. 1. *Sroki isčislyautsya časami,* **sutkami,** *mesyacami.* 2. *Srok, isčislyajemyj* **sutkami,** *istekaet v 24 časa poslednich* **sutok.** 3. *Pri zaderžanii srok isčislyaetsya s momenta faktičeskogo zaderžaniya.*[2]

---

[2] Emphasis mine, to highlight the frequency of using *sutki* as a timing unit in the statutory provisions (UPK RF); the variation of the form of highlighted *sutki* can be attributed to the noun declension system (available in Russian and unavailable in English), which prescribes nouns to change their cases in couple with their gender and number in communicating meaningful messages.

[The Russian Federation Penal Procedure Code as of 18.12.2001 No. 174-FZ (revised as of 12.11.2012) with amendments enacted from 21.11.2012. Article 128. Calculating the term of confinement. Para 1. The term of confinement shall be measured and accrued in hours, full days (*sutkami*), months. Para 2. The term of confinement calculated in full days (*sutkami*) shall expire at 12:00 p.m.[3] of the last day (*sutki*). Para 3. If detained at sight, the term of confinement shall be calculated from the time of actual detention.]

6 Another case of a lexical lacuna: How long does a *decade* – *dekada* last?

Conceptual lacunas are illustrated here by the word *decade*, a temporal unit that labels time-spans on the temporal continuum. A *decade* lasts a different length of time in English and Russian time-perception schema. When a *decade* is heard or read by English-speakers, it marks a timespan of ten (10) years, in contrast to the shorter duration it marks in Russian, amounting to less than a fortnight – ten (10) days.

Krysin's dictionary of foreign borrowings that have been assimilated into Russian suggests the following two definitions of *dekada*:
– *promežutok vremeny v desyat' dnej, tretiya čast' mesyaca* [a time span equalling ten days, a third part of a month]
– *desyatidnevnyj promežutok vremeny, posvyaschennyj kakomu-to obschestvennomu sobytiyu ili yavleniyu, obycno v literature ili iskusstve* [a ten-day period devoted to an event or phenomenon of public significance, usually in literature or arts]. (Krysin 2008: 222)

The significant discrepancy in meaning described above can be considered a conceptual lacuna between the temporal worldviews of English-speaking and Russian-speaking communities. One might initially assume that these words have the same meaning, seeing as both stem from the same Greek root δέκα or *deka*, indicating a multiple of ten (Webster's Unabridged 1989: 373).

As mentioned, a decade, measured on the English temporal continuum, denotes a period of ten (10) years, while in Russian *dekada* denotes a period of ten (10) days. Hence, the English expression "over the past decade" misleadingly appears to correspond with the Russian *v tečeniye poslednejch desyaty dnej* [over the last ten days], rather than *za posledneje desjatiletije* [over the

---
3 This time can also be marked as 00:00.

Table 2: Semantic distinctions between *decade* and *dekada* in English and Russian.

	decade – dekada		
English		**Russian**	
Denoted meaning(s)	**Time length** (quantifiable unit of time – a *year*)	**Denoted meaning(s)**	**Time length** (quantifiable unit of time – a *day*)
a decade *ten years* *a period of ten years* *a ten-year period* *desjatidnevnyj period*	10 years	*dekada* [a ten-day period] *desjat' dney* [ten days] [a ten-day period]	10 days

past ten years]. Such notable differences may give rise to cultural misunderstandings between the source-language speakers and target-language audience. Table 2 demonstrates the same-root words' deeper semantics.

For the legal translator, situations such as this necessitate cultural adaptation, challenging their techniques for achieving accuracy. It also implies or, rather, prescribes transposition and/or replacement, with optional explication of the translator's decision-making in their translator's notes. Strictly speaking, it is not so difficult to remember such divergence between the two words in question, but some contexts wherein a *decade* may be used can be filled with legal technicalities and thus become challenging to translate. However, there can also be self-obvious contexts where the meaning of a *decade* is so transparent that no hindrances will be encountered in its translation into Russian, as in the following example:

– In the past **decade**, the tobacco industry agreed to pay $246 billion to the states in a settlement of a lawsuit over the cost of smoking-related health care.[4]

No mistake is likely in translating this sentence, as the context makes explicit that the period in question is a ten-year period, totally excluding the Russian idea of a ten-day *dekada*.

---

4 Tobacco companies colluded to addict smokers, U.S. says. *The Baltimore Sun*. 22/09/2004. https://www.baltimoresun.com/news/bs-xpm-2004-09-22-0409220296-story.html. (accessed 16 January 2020)

6.1 Temporal properties of a *decade* exposed in translation practices

Despite the aforesaid, there are translation precedents to the contrary. The following case illustrates the intrinsic dissimilarity as it was handled by the UN Russian translators, according to the classical rules of ensuring equivalency in translation. One of the most significant efforts undertaken by UNESCO over the period of 1988–1997 was named the *World Decade of Cultural Development*. With the word *decade* in the title, the name of UNESCO's endeavour was rendered in Russian as *Vsemirnoye desyatiletiye razvitiya kultury* (UN Secretary-General, UNESCO Director-General Report 1989),[5] where a *decade* was translated as *desyatiletiye* preserving its English meaning – a "ten-year period" contracted into one word in Russian. That word-for-word rendering is undoubtedly correct. Unsurprisingly, there were so many pragmatically and situationally stipulated contextual details surrounding the big-scale event that Russian translators were unlikely to have made a mistake. By leaving the English meaning of that UNESCO endeavour in the Russian wording of the UN documents, the UN translators were able to be fully loyal to the source-language text, and their decision was professionally justifiable. Given the scale of the event and its announced official timeframe of 1988–1997, a mistake in the Russian translation would have been inexcusable.

Conclusions

This chapter has addressed some generalized approaches to cross-lingual conveyance of temporal meanings in the legal English-legal Russian language pair, by detailing instances of translation challenges in a variety of legal contexts. Principally, the research has exposed a complex interaction between the universal concept of time and its constituents across the temporal continuums employed by these two languages to express temporal meanings. Time, itself a universal human experience, is nevertheless experienced distinctly in different cultures, as manifested in various culture- and language-specific constructions of time.

The analysis above was based on several assumptions. Specifically, it was based on the assumption of the approximate nature of English-Russian translation correspondences between temporal lexis typically used in legal English

---

[5] UN Secretary-General, UNESCO Director-General Report 1989, World Decade of Cultural Development: 1989–1997. https://digitallibrary.un.org/record/64987?ln=ru. (accessed 17 November 2020).

and legal Russian to convey technical meanings within legal contexts. These were revealed through examples from culture- and language-specific contractual language, statutory and regulatory excerpts, as well as through definitions from encyclopaedic and specialized bilingual dictionaries of law.

Several conceptualizations of the Russian culture- and language-specific constructs visibly showcase pragmatic factors underlying the comprehension and respective meanings in translation. The above examples showed linguistic universals in their text-external timing, whereby event-timing is set by pragmatic factors, such as externalities, calamities, or other emergencies. Readings of language-specific temporal meanings across multiple legal contexts were demanding, in that they depended on the in-depth familiarity of the translator with the temporal meanings used habitually by native-speaking lawyers, even for such semantic universals as a 24-hour framework (*sutki*), and other temporal patterns preferred by language-speakers in each of the languages studied. The examples and charts conceptualizing a few patterns of the perception of *sutki* explicitly reveal that temporal meanings in legal texts are determined by text-external timing arrangements substantiated by events, such as the imposition of the state of emergency.

Dissimilar representation of semantic properties in temporal constructions provided insight into the root causes of the lack of homogeneity in conveying temporal meanings, which can be attributed to the inconsistency of the "English temporal space" with the "Russian temporal space." It is exactly the involvement of text-external timing that makes cross-lingual translation of temporal constructions conceptually complex. Text-external timing requires understanding culture-specific worldviews to arrive at valid translation correspondences. Russian temporal constructs such as *sutki* and *dekada*, exemplifying culture-specific lexis, do not have readily available equivalents in English but, rather, they have a number of approximately corresponding English lexical patterns used in similar situations. This lack of direct correspondences gives rise to lacunas across English-Russian legal languages.

In conclusion, there is every reason to complete the chapter by referring to the linguistic wisdom of Anna Wierzbicka, whose assertion we share and appreciate:

> In so far as everything we say we say in some language, so that even if we "translate" our thoughts from one language into another, we remain within the confines of a language.
> (Wierzbicka 1997: 23)

Time is a universal concept and a philosophical category expected to gradually evolve, specifically to serve the needs of the legal profession facing ongoing challenges in globalization. Hence research on temporal dimensions of legal knowledge and its cross-linguistic encoding may help other researchers to develop their insights in the expanding field of Legal Translation Studies.

References

Apresjan, Valentina J. 2012. The "Russian" attitude to time. In Luna Filipović & Katarzyna M. Jaszczolt (eds.), *Space and time in languages and cultures: Language, culture, and cognition*, 103–120. Amsterdam: John Benjamins.
Cheng, Le, Sin, King Kui & Anne Wagner (eds.). 2016. *The Ashgate handbook of legal translation*. New York: Routledge.
Filipović, Luna & Katarzyna M. Jaszczolt (eds.). 2012. *Space and time in languages and cultures: Language, culture, and cognition*. Amsterdam: John Benjamins.
Gladkova, Anna. 2012. Universals and specifics of "time" in Russian. In Luna Filipović & Katarzyna M. Jaszczolt (eds.), 167–188. *Space and time in languages and cultures: Language, culture, and cognition*. Amsterdam: John Benjamins.
Gotti, Maurizio & Christopher Williams (eds.). 2010. *Legal discourse across languages and cultures*. Bern: Peter Lang.
Huang, Minyao. 2012. Vagueness in event times: An epistemic solution. In Luna Filipović & Katarzyna M. Jaszczolt (eds.), 37–54. *Space and time in languages and cultures: Language, culture, and cognition*. Amsterdam: John Benjamins.
Giltrow, Janet & Dieter Stein (eds.). 2017. *The pragmatic turn in law: Inference and interpretation in legal discourse*. Berlin & Boston: De Gruyter.
Lewandowska-Tomaszczyk, Barbara (ed.). 2016. *Conceptualizations of time*. Amsterdam: John Benjamins.
Olsen, Frances, Alexander Lorz & Dieter Stein. 2008. *Law and language: Theory and society*. Düsseldorf: Düsseldorf University Press.
Olsen, Frances, Alexander Lorz & Dieter Stein, 2009. *Translation issues in language and law*. London: Palgrave Macmillan.
Pozzo, Barbara & Valentina Jacometti (eds). 2006. *Multilingualism and the harmonisation of European law*. The Hague: Kluwer Law International.
Ramos, Fernando Prieto. 2019. The use of corpora in legal and institutional translation studies: Directions and applications. *Translation Spaces* 8 (1). 1–11.
Ramos, Fernando Prieto. 2021. Translating legal terminology and phraseology: Between intersystemic incongruity and multilingual harmonization. *Perspectives: Studies in Translation Theory and Practice* 29 (2). 175–183.
Robertson, Colin D. 2016. *Multilingual law: A framework for analysis and understanding*. London: Routledge.
Šarčević, Susan. 2000. *New approach to legal translation*. The Hague: Kluwer Law International.
Stepanov, Yuri S. 1997. *Konstanty: Slovar' russkoj kultury: Opyt issledovanyja* [Constants: Dictionary of Russian culture: Research experiences]. Moscow: Skola Yazyki russkoj kultury.
Tiersma, Peter M. 2001. Textualizing the law. *International Journal of Speech, Language and the Law* 8 (2). 73–92.
Tiersma Peter M. & Lawrence Solan (eds.). 2012. *The Oxford handbook of language and law*. Oxford: Oxford University Press.
Vlasenko, Svetlana. 2006. *Contract law: Professional translation practices in the English-Russian language pair*. Moscow: Wolters Kluwer.

Vlasenko, Svetlana V. 2016a. Minimal unit of legal translation vs minimal unit of thought. In Le Cheng, King Kui Sin & Anne Wagner (eds.), *The Ashgate handbook of legal translation*, 89–120. London: Routledge.

Vlasenko, Svetlana V. 2016b. Where 'fiscal' cannot mean 'financial': A case study at the crossroads of legal and public-service translation taxonomies. *New Voices in Translation Studies* 14. 46–73.

Vlasenko, Svetlana V. 2017. Legal translation pragmatics: Legal meaning as text-external convention – the case of 'chattels'. In Janet Giltrow & Dieter Stein (eds.), *The pragmatic turn in law: Inference and interpretation in legal discourse*, 249–286. Berlin & Boston: De Gruyter.

Wierzbicka, Anna. 1996. *Semantics: Primes and universals*. Oxford: Oxford University Press.

Wierzbicka, Anna. 1997. *Understanding cultures through their key words: English, Russian, Polish, German, Japanese*. New York and Oxford: Oxford University Press.

Wierzbicka, Anna. 2002. Russian cultural scripts: The theory of cultural scripts and its applications. *Ethos* 30 (4). 401–432.

Legal sources

Child marriages 2013: Child marriages: 39 000 every day / World Health Organization. Report. https://www.who.int/mediacentre/news/releases/2013/child_marriage_20130307/en/ (accessed 15 November 2020).

Garant 2020: Garant+ Russian Legal Data Base. http://base.garant.ru/ (accessed 20 November 2020).

Looking ahead towards 2030: Eliminating child marriage through a decade of action / UNICEF. September 2020. https://data.unicef.org/resources/looking-ahead-towards-2030-eliminating-child-marriage-through-a-decade-of-action/ (accessed 15 November 2020).

RF Supreme Court Resolution 2012: Postanovlenyje Plenuma Verchovnogo Suda RF ot 01.02.2011 (red. ot 09.02.2012) "O sudebnoj praktike primenenyja zakonodatel'stva, reglamentiruyuščego osobennosti ugolovnoy otvetstvennosti i nakazanyja nesoveršennoletnych" [The Russian Federation Supreme Court Plenary Session Resolution N1 as of 01.02.2012 (revision as of 09.02.2012) "On Judicial Practices for Enforcing Legislation Regulating Criminal Liability and Penalty for Minors", para 2(5)]. In *Rossyjskaya Gazeta*, 29. dd. 11.02.2011.

UN Secretary-General, UNESCO Director-General Report 1989, World Decade of Cultural Development: 1989–1997. https://digitallibrary.un.org/record/64987?ln=ru (accessed 17 November 2020).

UPK RF 2012: Ugolovno-prosessual'nyj kodeks Rossijskoy Federacii ot 18.12.2001 N 174-FZ (red. ot 12.11.2012) (s izm. i dop., vstupayuščimi v silu s 21.11.2012). Statjya 128. Isčislenyje sroka. [The Russian Federation Penal Procedure Code as of 18.12.2001 No. 174-FZ (revised as of 12.11.2012) (with amendments to be enacted from 21.11.2012). Article 128. Calculating the term of confinement].

Dictionaries

ABBYY 2014: *ABBYY Lingvo x6.0*. Electronic multilingual dictionary. ABBYY Software Ltd., 2014. https://www.lingvo.ru/ (accessed 27 November 2020).
Faekov, Vladimir. 2011. *New finance dictionary*, 2nd edn., 2 vols. Vol. 1: English-Russian. Vol. 2: Russian-English. Moscow: International Relations Publishing House.
Garner, Bryan A. 1987. *A dictionary of modern legal usage*. Oxford: Oxford University Press.
Garner, Brian A. (ed.). 2009. *Black's law dictionary*, 9th edn. St. Paul: Thomson Reuters.
Krysin, Leonid. 2008. *Illustrirovannyj tolkovyj slovar' inostrannych slov* [illustrated explanatory dictionary of foreign borrowings]. Moscow: Eksmo.
Oxford Russian 2000: Thompson, Della (ed.). 2000. *The Oxford Russian Dictionary*. 3rd edn. Oxford: Oxford University Press.
Pivovar, A. G. (ed.). 2003. *Great financial and economic dictionary*, 2nd edn. Moscow: Ekzamen.
Webster's Unabridged 1989: *Webster's Encyclopaedic Unabridged Dictionary of the English Language*. New York: Portland House.
Yefremova, Tatyana. 2005. *Tolkovyj slovar' slovoobrazovatel'nych edinic russkogo jazyka* [Explanatory dictionary of word-formation units in the Russian language], 2nd edn., rev. Moscow: Astrel'-AST.

Media sources

Tobacco companies colluded to addict smokers, U.S. says. *The Baltimore Sun* 22/09/2004. https://www.baltimoresun.com/news/bs-xpm-2004-09-22-0409220296-story.html (accessed 16 January 2020).

Subject index

analogy 11, 109, 109 n7, 112, 115–121, 117 n11, 117 n12, 123, 125
announcement of purpose 88, 93
argumentation 21, 57, 58, 60–62, 64–66, 68, 68 n9, 71
Ariel, Mira 12–14, 12 n7, 131, 138–147, 149–153, 155, 156, 159
– "that is to say" test (faithful-report test) 11–14, 142–144, 155, 156
art, artistic expression 1–6, 1 n1, 17–31, 57–76, 92, 100, 183
– art, as exceptional speech 18, 19
– rap/hip hop 2, 23–30, 24 n5
Austin, John L. 36, 134
attitudinal model 58

background assumptions 149, 150
Baron, Dennis 81, 82, 94
Bourdieu, Pierre 18
Brown, Penelope & Stephen C. Levinson 5, 42, 43, 44, 46

canons of interpretation 12, 93, 132
Carston, Robyn 134, 139, 140, 153, 154, 156–160, 159 n32
censorship 17–19, 21
cognitive linguistics 10, 107–110, 107 n1, 114–117, 123–126, 125 n16, 131, 134, 135, 140, 153, 155, 156, 158
– cognition and categorization 110, 115–117
composite audience 57
conceptual blending 124–126
Conceptual Metaphor Theory 10, 11, 108–111, 115, 117, 119, 120, 123, 124
conventionality / cueing function 111, 111 n7, 116, 120, 121, 123
corpus linguistics, corpora 1, 7, 8, 9, 18, 21, 29, 59, 68 n10, 77–107, 123, 126
– CEME (Corpus of Early Modern English) 9, 80, 81, 83
– COCA (Corpus of Contemporary American English) 68 n10

– COFEA (Corpus of Founding Era American English) 9, 80, 81, 83–86, 88, 93, 94, 95–106
Conversation Analysis 4, 44–49
critical discussion 61, 62, 64, 70, 71

dialogism 69, 69 n11
discoursal status 12, 13, 139, 140–144, 147, 149, 150, 152
distanciation 1–3, 18, 22–29
Dworkin, Ronald 71

embodiment 110, 111, 114
emotion 31, 34, 36, 38, 43, 47, 52, 57, 58, 65
– emotive language 58, 65
empircal research 108, 113, 117, 118, 126, 168
evaluation 22, 24, 37, 57, 58, 59, 59 n3, 59 n4, 59 n5, 60, 61, 62, 64, 65, 66, 67, 68
– evaluative language 58, 60, 61, 62, 65, 67, 68, 69 n11
evidence, language as judicial evidence 4, 24 n5, 32, 35, 36, 37, 42, 53, 63, 67, 69, 108, 113, 181
evidence, language as scientific evidence 10, 68 n10, 88, 108, 109, 114, 117, 123, 124, 125
evidence, language as pragmatic evidence 133
experimental linguistics 10, 11, 107–130, 154, 155, 160, 160 n33
explicature 12, 13, 132, 139, 140, 142–144, 146–149, 153–156, 158, 159
– *ad hoc* concept 156, 158, 158 n.31
– provisional explicature 146, 147
explicit meaning 133, 138–141

Fauconnier, Gilles & Eve Sweetser 125, 126
free exercise clause 6, 58, 63, 70, 71, 72
freedom of art (*Liberté de création*) 1, 3, 4, 5, 17–30

Subject index

freedom of choice and action 44, 45, 47, 52
freedom of speech 6, 17–30, 19, 20, 21, 57–76, 62
folk linguistics 8 n6, 22, 24
force dynamics 110
forensic linguistics 32, 37, 107, 165
frame 11, 36, 39, 40, 52, 62, 81, 82, 88, 107–130, 169, 176, 177, 178

grammar 7–10, 8 n6, 60, 77–106
- absolute phrase 8, 9, 77–91, 93–106
- adverbial clause 88
- conditional clause 171
- independent clause 8, 9, 77, 78, 80–82, 85, 87, 88
- operative clause 8, 9, 77, 82, 86–88, 93, 94
- prefatory clause 8, 9, 77–80, 82, 86–88, 90, 91, 93, 94
- purpose clause 94
Grice, Paul 3, 10, 12, 36, 47, 47 n7, 48, 48 n8, 50, 50 n9, 134, 134 n6, 139, 140, 153, 154, 159
- Gricean cooperative principle 35, 36, 47, 47 n7, 50, 51, 134
- Gricean maxims 47, 48, 48 n8, 50, 50 n9, 134, 153, 155, 157, 159
- neo-Gricean pragmatics 132–134, 134 n7, 138–140, 153–158, 155 n25, 159, 159 n32, 160
- post-Gricean pragmatics 132–134, 138–140, 153–155, 159, 160

Hart, H.L.A. 115
hate speech 1, 20
Horn, Laurence 133, 134, 134 n7, 139, 153–158, 155 n25, 159 n32, 160
- division of pragmatic labor 153, 155
- Horn scale 154, 156–158
- hypernym 156–158
- hyponym 156–158
- Q Principle 153, 154, 157
- R Principle 153–155, 157
Hunter, Dan 117–120, 122
Hymes, Dell 4, 38, 39

implicature 12, 13, 47, 48, 50–53, 139, 140, 144–149, 153–157
- conversational implicature 12, 13, 47, 48, 50, 51, 52, 53, 147–149
- particularised conversational implicature 147–149
- strong implicature 13, 144–146
indirectness 3, 4, 5, 13, 32, 133, 144–147, 149, 150, 153
inference 3, 12, 23, 32, 47, 49–51, 110, 131, 133, 134, 138–143, 150–153, 155–160, 166, 174
- analogical inference 166, 174
- explicated inference 12, 140, 142, 153, 155, 156
- pragmatic inference 139, 140, 143, 153, 156, 158, 160
- truth-compatible inferences 150–152
international law 13, 20, 131–164, 167

judicial decision 2, 6, 7, 21–26, 57–60, 61 n7, 62, 64, 65, 66, 68 n9, 70, 71, 73, 77, 80, 89, 108, 115, 116, 117, 118, 120, 123–125, 148, 151, 151 n22, 166

lacunas, conceptual 14, 166, 174, 183, 186
lacunas, deep 174
lacunas, lexical 14, 166, 173, 174, 177, 178, 183, 186
Lakoff, George 109, 110
Lakoff, Robin 43
legal argumentation 21, 57–62, 61 n7, 64–66, 68, 68 n9, 70, 71
legal interpretation 2, 4, 5, 8, 11–14, 17, 20–24, 24 n5, 27, 35, 36, 39, 65, 80, 82, 93, 94, 115, 116, 125, 126 n17, 131–164, 165–167
legal precedent 11, 63, 109 n3, 113, 114, 114 n8, 117, 117 n12, 118, 119, 120–122, 124, 125, 185
legal reasoning 58, 62, 63, 66, 67, 68, 69, 72, 107 n1, 109 n3, 111, 112, 116 n10, 117 n12, 118, 125
Likert scale 113, 121
literal meaning 2–4, 13, 17, 50, 113, 120, 121, 144–147

metaphor 10, 11, 107–130
minority rights 2, 17, 23, 57

ordinary meaning 4, 6, 7, 11, 12, 14, 80, 107, 126, 126 n17, 126 n18, 132, 136–138, 136 n9, 141, 144, 152, 157, 158 n31, 159
originalism 80, 93

Politeness Theory 4, 5, 36, 42–44, 46, 47
– impoliteness 36, 42, 42 n5, 43, 47
– face-threatening acts 5, 36, 42–47, 51, 52
– negative face 36, 42, 43, 44, 45, 46, 47, 52, 53
– positive face 42, 43, 46, 47, 52, 53
pragmatic determinants 168, 169, 181
pragma-dialectical theory of argumentation 58, 60, 61, 61 n7, 70, 61, 62, 64, 64, 66, 66 n8, 71, 73
prefatory-materials canon 86, 93, 94

Relevance Theory 3, 8, 12, 131–163

Scalia, Justice Antonin 7–10, 60, 80–82, 86–88, 93, 94
Searle, John R. 36, 134
Second Amendment 7, 8, 77–83, 85–88, 93, 94
semantics 14, 48, 77, 82, 83, 87, 88, 93, 126, 132, 133, 136, 138–141, 143, 154, 157, 158 n31, 159, 165, 168, 169, 174, 176, 177, 182, 184, 186
– semantics and pragmatics 126, 133, 136, 138, 168

sexual misconduct 4, 31–56
– *quid pro quo* sexual harassment 4, 31–33, 47, 51–53
sexual orientation 66, 67, 71
Speech Act 4, 35, 36, 61, 134, 134 n5, 142
Speech Event 4, 5, 35, 36, 38, 39, 40, 42, 43, 47, 51, 52
Sperber, Dan & Deirdre Wilson 3, 12, 134, 135, 140, 142
Supreme Court (United States) 6, 57, 58, 64, 68 n9, 70, 71, 72, 77, 80
Supreme Court (Poland) 57 n1
Supreme Court (Russia) 182
stance-taking 59
strategic manoeuvring 61, 63, 70, 71

Tiersma, Peter M. 107, 108 n2, 165, 177
time/temporality 14, 15, 77, 83, 88, 108, 110, 165–189
tort and insurance law 123, 155, 156, 172, 181
translation 14, 15, 16, 22, 125, 165–189
treaties, treaty interpretation 11, 12, 13, 14, 131, 132, 132 n1, 132 n2, 133, 136, 137, 138, 141, 144, 145, 147, 148, 149, 150, 152, 159
truth-conditions 140

Vienna Convention on the Law of Treaties 132, 136–138

Wittgenstein, Ludwig 115, 133

www.ingramcontent.com/pod-product-compliance
Lightning Source LLC
Chambersburg PA
CBHW031432150426
43191CB00006B/480